THE ENERGY PREDICAMENT:

EXPLORING THE REALITIES BEHIND MODERN AND FUTURE ENERGY SOLUTIONS FOR CLIMATE CHANGE

JEREMIAH CUTRIGHT

DORRANCE PUBLISHING CO
EST. 1920
PITTSBURGH, PENNSYLVANIA 15238

Dorrance Publishing Co
585 Alpha Drive
Pittsburgh, PA 15238
Visit our website at *www.dorrancebookstore.com*

ISBN: 978-1-6393-7048-1
eISBN: 978-1-6393-7837-1

THE
ENERGY
PREDICAMENT:

EXPLORING THE REALITIES BEHIND MODERN

AND FUTURE ENERGY SOLUTIONS

FOR CLIMATE CHANGE

Table of Contents

A Note on Units:

Energy/Electricity:

The Watt (W) is a unit of power. Units can also be expressed in units of energy output. For example, 1 Watt-hour(Wh) is the same as running 1 Watt of power per second, for an hour. Or, a watt of power over an hour has the same amount of energy as a Watt-hour. Typically, units of *power* will be used in terms of electric capacity (that is, the energy that could be generated) and units of *energy* will be used in terms of electricity actually produced. For reference, most power plants have between 800-1,200 MW of electric capacity power, and the average American home will use 1,000 kWh of electricity per month.

1 Watt over an hour= 1 Watt-hour (Wh)
1,000 W=1 Kilowatt (kW)
1,000 kW=1 Megawatt (MW)
1,000 MW=1 Gigawatt (GW)
1,000 GW=1 Terawatt (TW)
In 2018 the world consumed 25,000 TWh of electricity and had an electric capacity of 9,700 GW.

Weight:

In most instances, "tons" refers to metric tons (also commonly represented as "tonnes"). 1 metric ton=1.1 US tons. The world emits around 51 Gt of anthropogenic CO_2 equivalent emissions each year (as of this writing in 2021).

1,000 tonnes=1 Kiloton (kt)
1,000 kt=1 Megaton (Mt)
1,000 Mt=1 Gigaton (Gt)

Preface

The main objective of this book is to provide readers with the knowledge that they require in order to make fair, reasoned, and educated conclusions on subjects of energy concern. While the underlying motive in the transition of energies is due to climate change, this book will not heavily focus on these factors—though they are discussed lightly throughout and at the end of the book. The realities of the issues we face are frequently represented to the public through mediums that cherry pick information or simply conduct in disinformation. The primary subject matter of this book is the discussion of common misconceptions in the energy sector. For example, renewable energies have drawbacks that are ignored, and nuclear energy has positives that are disregarded in the same fashion. Because of these discrepancies, this book is particularly critical on renewables and equivalently lenient on nuclear energy, though the opposite viewpoints are also given attention. I want to be clear that despite this, it is important to keep in mind throughout this reading that renewables (particularly wind and solar) have colossal societal benefits and by no means should their research, development, or implementation be halted. If anything, the opposite is true in order to solve the issues that this book brings attention to.

The following chapters will provide an (frequently data heavy) explanation of various energy misconceptions and explain them so that you (the reader) can be equipped with a level-headed view of energy realities and also have the numbers to back it up. It is one thing to say fossil fuels are dangerous. It is another to explain, with empirical data, that they are responsible for the deaths of countless *millions*. The cornucopia of data thrown around may confuse some readers at times during this book, but I do my best to explain the process and represent any scratchwork done that may be useful—either through page footnotes or notes in the back of the book. For this same reasoning, I have decided to include an extensive list of citations at the back of this book, to the point of redundancy in many cases, so that any readers are able to cross-check any information laid out in this writing.

While this writing is based on a compendium of scientific research and available data, the objective throughout is to convey this information in a way that is simple enough for any reader, of any background, to understand, yet nuanced enough to not lose the important information. Because of this, many sources cited do not relate to hard scientific data (although many do) but will reference back to resources such as fact sheets, documentaries, videos, and other *free publicly available information* that can be easily accessed, understood, and shared. Also because of this, some information is simplified for clarity throughout, though this is also noted in the text where it occurs most prevalently. Extensive effort was spent to find free, publicly available information that is also reliable in order for accessibility, but there are a non-zero number of citations that will be hidden behind pay walls (especially regarding published scientific studies in prevalent journals).

This writing was heavily influenced by research coming from a multitude of international organizations, governmental organizations, private institutions, and independent studies and persons, so as to minimize conformation bias and gather opposing viewpoints. In saying that, it was most heavily influenced by various reports from the United Nations' Intergovernmental Panel on Climate Change—specifically their *Fifth Assessment Report*, but also the *Global Warming of 1.5°C* and *Climate Change and Land* special reports; the International Energy Agency's *World Energy Outlook, Tracking Clean Energy Progress, Global Energy Review* and *Sustainable Development Scenario* reports; various other organizations within the United Nations; various departments of the US government—specifically the Energy Information Agency, Department of Energy, and Environmental Protection Agency; and the vast amounts of information available on the websites Our World in Data, Statista, and Worldometer. Though not directly discussed in the text, I have done my best to include impacts that relate to the UN's Sustainable Development Goals, and therefore a wide range of impacts and concerns are discussed.

Some persons may not find some of these sources acceptable or reliable. While there are flaws in organizations such as the United Nations, the reality is that these systems are comprised of thousands of leading scientists, economists, and otherwise professionals and represent the most prestigious, volumetric, and respected peer-reviewed information coming from scientific research across the globe. And, as I mentioned previously, this information is frequently compared to other sources with no or little relations so as to cross-check information.

The debate whether climate change is happening or not is now largely a waste of time and energy and so "debunking" climate change denial reasoning will not be a main focus of this book. When 97% of scientists agree that our planet is warming, that is as close to scientific consensus as you will get. This does not mean that the 3% are smarter or less intelligent than the other 97%; it simply means that their data has come to a different conclusion than that of others (in some cases there may be external factors such as personal bias, conflicts of interest, or selective funding involved). If 97% of people told you it would be a bad idea to jump off a bridge, does that mean you should listen to the 3% who say it would be good for you? Sometimes, you have to accept that not everyone will agree on something and then move on to fixing the problem without them. Because of this, this book is not about debunking climate change denial, but will focus in on some aspects of this towards the end, just to give the average reader a sense of the scale of agreement about this topic within the scientific community.

Many topics in this book will be useful tools far into the future, but because of the constantly evolving science, economics, and political motives to adapt our energy intensive world to be more sustainable, some parts of this book will also undoubtedly become increasingly less relevant as time goes forward. Objective dates will pass, situations will change, and innovations will occur at exponential rates. Even over the two years I have spent working on this book, many aspects changed, and various sections (even chapters) had to be heavily edited or rewritten as more information came out and situations around the globe evolved at light speed. I wish that the single most important takeaway any readers get from this writing will be to constantly question the world around them and to be able to better recognize misinformation and disinformation when it is near. Finally, when discussing energy (and anything else in life for that matter), it is important to be skeptical, but not ignorant of reality.

Introduction

The Modern Situation
Defining The Energy Predicament

For all of human history we have hardly become more advanced in terms of how we actually produce energy. Until the late 20th century, the same technique was applied to most systems of energy production—we simply set things on fire. At the earliest, cavemen discovered fire 300-400,000 years ago.[1] They learned how to tame this source of pure energy to aid them in life. What we now consider simple parts of life were revolutionary to them. To be able to provide heat in the middle of the night so you don't freeze to death. To have a source of light to see in the dark. To cook food so it was safer to eat. This was all done by just lighting wood and animal products on fire. Fast forward to the 1870s and we began developing the first effective electric light bulbs, no longer requiring the burning of wood or animal oils to provide light. We then created ways to use electricity to create heat and to cook our food as well. At this moment in time, we realized we needed a way to produce electricity for mechanisms on an industrial and commercial scale, quickly and cheaply. While hydro and wind were also obviously widely used, their applications were very limited and so when fossil fuels came around, they rapidly overtook the marketplace.

Coal quickly took over as the main source of power in the world. Again, the principle behind generating energy with coal was the same as beforehand. We would bring in coal to a factory and light it on fire. The only difference this time was that we now had to create electricity. To do this, all that needed to be done was to use the heat from the coal to steam water which then would turn a turbine and then a generator and eventually create functional electricity. This method—typically referred to as a steam turbine—turned the potential energy of the coal into heat which caused the steam to become mechanical energy and finally electricity. This can seem to be a complex way of producing power, but the original principle remained the same—we light stuff on fire.

Even once oil and natural gas came along, this system did not change. (Even in circumstances where fire is not being used, this method of spinning a turbine has become prominent. Nuclear power uses a steam turbine but does not require any material being set on fire. Hydroelectric power spins turbines with the flow of water, and wind power spins a turbine with the wind. Today, the only mainstream form of energy that is inherently different is photovoltaic solar panels—although concentrated solar power still uses a steam turbine method.)

With the industrial revolution having led to an extremely prosperous time (relative to the rest of human history), there was a much greater transition into scientific research to continue this idea of infinite growth. This newfound pressure for scientific knowledge is what propelled the world into what we can call home today. Because of great discoveries in the sciences and manufacturing, an economic, social, and technological boom quickly followed. Sadly, it was in this time period of human history where humanity started to lose itself. Even in the late 1800s, scientific research began to pull through about how bad the burning of fossil fuels could be for the environment and human health, and how carbon-dioxide (CO_2) emissions from these practices might warm the earth. Svante Arrhenius[1]*, a well-renowned Swedish scientist who would later go on to win the Nobel Prize in chemistry in 1903, was one of the first to describe the greenhouse effect in detail. In 1896, he wrote in an article for the Philosophical Magazine and Journal of Science titled "On the Influence of Carbonic Acid in the Air upon the Temperature of the Ground": "A simple calculation shows that the temperature in the arctic regions would rise about 8° to 9°C., if the carbonic acid increased to 2.5 or 3 times its present value. In order to get the temperature of the ice age between the 40th and 50th parallels, the carbonic acid in the air should sink to 0.62-0.55 of its present value (lowering of temperature 4°-5°C.)."[2]**

It is pointed out further in this article the effects of coal burning and its direct relationship to carbonic acid (which is formed when CO_2

---

* Who, coincidentally, is distantly related to the now-popular youth climate activist Greta Thunberg.
** Arrhenius's predictions would go on to be a solid evidence that GHGs could warm the planet, but his predictions on how much they would warm the planet were off. Although CO2 concentrations have increased far quicker than he predicted, the temperature has increased by less.

dissolves in water; $CO_2+H2O=H2CO_3$). As research continued, we became increasingly aware of the potential risks by the mid 1900s, but unfortunately, this information was largely ignored, and we were set on a path of destruction. Arrhenius was not the only voice to speak up about this topic. While many during the latter half of last century considered the science of anthropogenic (man-made) climate changes "not settled," the reality was far from this. Scientific consensus on the likelihood of anthropogenic climatic influences was strong into the 1980s, and today there is as close to a scientific consensus that you will get in any subject. (See chapter 17.)

Had science been listened to earlier on, we would have been able to avoid much of the environmental and social damage that we are witnessing today. Now, global anthropogenic climate change is rapidly becoming one of the most studied scientific topics in history as we try to understand just how screwed we are.

It wasn't until overwhelming research finally started pouring through that these claims would finally start to be heard by governments and large energy providers. Even though renewable energies such as solar had been hypothesized and prototyped as early as the 1800s it has only been relatively recently that they started to gain momentum as viable forms of power. Solar, for example, was invested in and researched by the US government as early as the 1970s but has not gained serious traction until recently.[3]*

Nowadays, we are starting to move away from fossil fuels and towards better sources of energy. Renewable energies are coming to the world in a big way. In the US alone there are now over 2 million solar installations.[4] With the addition of wind, geothermal, nuclear and hydropower, the world is transitioning to not only a cleaner world, but also a more diverse and de-centralized one. As we will see throughout this book, these technologies have massive implications beyond just the mitigation of climate change. More solutions, however, can also mean more problems. This is especially true of problems that were not anticipated.

Renewables almost certainly will not be able to solve the crisis of climate change on their own. Renewables like solar and wind energy

---

* While much of this reasoning is because of political motivation, the technological capabilities back then were not as good as today. The possibility for efficient and cheap commercial-scale solar power has not been feasible until this century.

are seen as 100% clean by most and so they seem like the obvious choice of power. However, the things that are almost always overlooked are the obvious downsides to solar and wind, whereas the negatives to sources such as nuclear are blown way out of proportion. Natural gas power could help us to reduce our carbon footprints but will not get rid of them, so there is continuing debate whether it is worth our time and energy. Considering that US CO_2 emissions peaked all the way back in 2007 and have fallen significantly since then due to the replacement of coal with natural gas, there is a serious argument to be made for natural gas as a transitionary power source. Unfortunately, because of these unforeseen issues and conflicting ideologies, we have set for ourselves two distinct mainstream paths, each of which could prove disastrous.

On the first path, we ignore the threats of climate change and continue our use of fossil fuels in the name of infinite economic growth and in order to bring developing countries out of poverty. This however will obviously bring about catastrophic consequences through global warming (see chapter 19). When this path recognizes there is an issue (seldom from some parties), the solutions brought forward are usually established sources of energy such as nuclear and hydroelectric, but completely ignore the massive benefits that could be had from wind and solar. This path hopes to rescue the economy in the short term while putting us at massive risk from climate change in the long term.

On the second path, severe and drastic measures to transition to an economy that prioritizes environmental wellbeing over everything else (primarily through the use of solar and wind power) are brought to the forefront. While this is nice in theory, it could show to be damaging in practice. Solar and wind have potential for economic growth and environmental health, but they also have potential to wreak havoc on economies and threaten the wellbeing of a society in unforeseen ways if they are not handled properly and responsibly. This effort is only damaged further by the fact that more stable forms of low-carbon power generation (such as nuclear and hydroelectric) are not seen as viable because of the perceived environmental concerns that are not always based on accurate information. This path hopes to save the world from climate change, while also ignoring various aspects of reality that threaten the global economy.

When the efforts and funding behind stopping climate change are put into situations that are guided by opinions, and not the actual facts of how to help, these problems are only amplified. The improper directory of resources to unrealistic solutions will not only waste precious time and money but will also pose a threat to the economic and geopolitical stability of countries all over the globe.

Our predicament can therefore be defined as thus: Following either path could lead to the destruction the other hopes to avoid. The only way for us to avoid both threats is to compromise, accepting that both paths are right, and both paths are also wrong. To maximize benefits and minimize damages, we must create a new path altogether.

These intersecting problems of energy solutions put the world at a crossroads between solutions that are great in theory and solutions that are decent but already here. We need solutions now but can't agree on what the solutions are or (in the case of the US) if there is even a problem to begin with.* Solar and wind are great in theory but are not ready—in some cases—for the situations they are being applied to, and yet the US Republican Party largely refuses to acknowledge that they have any benefits at all. Nuclear power is our only option for carbon-free electric that can be applied nearly anywhere on the planet and is not intermittent, and yet the US Democratic Party and many environmentalists feverously oppose any adoption of it and actively seek the dismantlement of the technology. We do not have all of the solutions we need yet (this is most heavily highlighted in section III), but progress is severely lagging behind where it needs to be because of this political stalemate. Additionally, many are tackling the "low-hanging fruit" of solving climate change (such as electric generation and road transportation), but fewer are thinking about the hard problems (such as industrial activities, air travel, and carbon capture technology).

Due to the large gap in practical information being circulated, this effect is only intensified when we elect politicians who dispense false claims or only focus on the easy stuff, further pushing improper etiquette. It is often not their fault; they are just as susceptible as

---

* Spoiler alert: there is.

everyone else to the misunderstanding of many factors. But still, the situation only snowballs itself. That is why this book will be focused not on the parts of energy generation and climate change that are widely discussed, but the things that nobody wants to talk about and the issues that are commonly misunderstood and often blatantly lied about. This book will not focus primarily on climate changes effects as there is already a cornucopia of knowledge being circulated (although chapter 19 will talk about it) but will instead assume that this is a problem the world must face and discuss how to face it in a way that will not only solve the problem in the most effective way, but will simultaneously lift millions out of poverty, make the world more peaceful and plentiful, and invigorate the stagnating world economy.

Climate change gives us a rare opportunity to help the world as a collective species. We have the chance to push the human race ahead in prosperity and technology at light speed if we can solve this issue the right way. However, if we go about solving this problem the wrong way it will certainly have massively negative effects on the same factors that could benefit. Additionally, this book will not focus on politics (although specific political actions may be discussed, they are not reviewed outside of this book's scope) or policymaking for the reasoning that they are largely not helpful for understanding the content of this book.

Most of this book will focus on how electricity is made. This is not simply because it is the biggest single source of carbon emissions, but because decarbonizing electric generation will allow us to electrify—and therefore decarbonize—other sectors (through technologies such as electric cars). In fact, an Intergovernmental Panel on Climate Change (IPCC) report found that electricity and heat production was only responsible for 25% of global Greenhouse Gas (GHG) emissions, while agriculture, forestry, and other land use was responsible for 24%, industry 21%, and transportation 14%.[5] Throughout these chapters, "electricity" and "energy" will largely be used interchangeably for ease of reading, but keep in mind they are not necessarily the same. Technically, "energy" refers to not only electricity but also fire, calories, heat, and various other subjects. Electricity is simply one of many forms of energy available to us today.

Section I:

A Challenging Utopia

Chapter 1

The Solutions We Want

Introduction to Renewable Power

Many are quick to blame renewables for any problems possible. The Texas blackouts in early 2021 that were caused by unusually cold weather are a prime case study of this. Immediately, the claim of "wind turbines freezing" captured headlines around the US and many in the Republican Party attempted to take this opportunity to speak out in opposition of renewable energy. These claims were nothing but deeply flawed. The idea of wind turbines freezing (which, to be clear, did happen) being a fatal flaw in the technology is an absurd and short-sighted thought process. Some of the countries in the world with the highest quantities of wind power (Denmark, Sweden, Norway, Germany, the UK)[1,2] are indeed, some of the coldest. Texas simply decided, despite various federal warnings,[3,4] to not incorporate protective measures to their wind turbines and other infrastructure that would have protected them against unusually cold weather. Additionally, keep in mind that the 2021 blackouts were not the first to ever happen to Texas. The state experienced similar events in 1983, 1989, and 2011.[5] Furthermore, wind was not the only culprit, but all forms of energy production were affected due to the lack of preparedness by the state. Based on total power lost, natural gas was the worst offender.

Situations like this are frequent with both renewable and nuclear energies. In reality, blaming wind turbines on freezing during this disaster is like blaming the Fukushima incident on nuclear energy. They both resulted because of corporate greed, as we will see later. In both of these examples, something significant but unrelated to the fundamentals of the power source is taken out of context so that it can be used against the technology and those who support it. It is also important to remember that not all renewable sources are intermittent. Hydroelectric power, for example, has provided reliable and consistent baseload power for longer than any other form of electric generation (see chapter 4). Geothermal power has limited applications but is even more of a consistent power source than hydroelectric and fossil fuels (see chapter 5).

Solar and wind are intermittent but have seen immense cost declines in recent history. In the US, construction costs for photovoltaic solar fell by 50% between 2013 and 2018, and wind construction costs over the same time period fell by 27%.[6] From 2010-2019, overall photovoltaic solar costs fell by 82%, concentrated solar (also called thermal solar) fell by 47%, onshore wind by 40%, and offshore wind by 29%.[7] This does not mean they are now economic in every application, but the trends show that they are becoming increasingly competitive and soon will be competitive in most cases. The cost reductions being seen in the renewable energy sector are extremely impressive.

Still, one main barrier remains for solar and wind. This is energy storage. Solar and wind can produce massive amounts of power when they are active, but without effective ways to store this power for later, they will not be able to supply us with a reliable power source. (Another possible solution for this is to include long-distance transmission capabilities so that, for example, excess solar power from a sunny region can be efficiently transported to another that is cloudy. In reality, both of these solutions are desired in order to make a quick and effective transition to solar and wind.)

Energy Storage

There are several countries where solar and wind are making great progress. As great as this is, many other larger countries aren't able to meet the energy demand in large cities with these because when the wind dies out or the sun stops shining, we drain the batteries storing energy far too quickly. This is also a problem that is bigger than just storing enough energy overnight, for 6-8 hours each day. This is a problem that requires a solution where we can store enough energy for at least one week, in the case of major weather anomalies (like Texas in early 2021) where there are large decreases in the amount of sunshine or wind that is available or being captured. In the case of larger lulls in energy production opportunities, this can become a serious national security concern. What if smoke from a wildfire blocks out the sun in a large region for weeks, or worse yet, if a major volcano erupts and clouds out the sun over the planet for a year (such as was the case with Mount Tambora—whose eruption caused "The year without a summer" in 1816)? What if there is an abnormally low amount of wind for a couple months? What if global warming changes wind currents and installed wind power slowly becomes less efficient?[8] The list goes on, and on. There are those that think, as a matter of national security, that because of these unlikely but severe risks, solar and wind should not be considered.

In lieu of the more existential of these concerns, energy storage is researched more and more every day. As with the energy production makeup of the future, there will almost certainly need to be several solutions for storage. Additionally, it is important to understand that all energy storage systems are actually energy consumers. Since they are not producing the energy themselves, there is inherently a non-zero amount of energy loss that can be attributed to each type of storage technique, though these losses vary widely. An overview of various types of energy storages will be helpful to understand for the coming chapters.

Lithium-Ion Batteries. What is likely now the most popularized form of energy storage is Lithium-Ion (Li-Ion) battery storage. This is the same type of battery that is in your phone and possibly powers your car. The most popular Li-Ion storage facility is undoubtedly the Hornsdale Power Reserve in Australia. In 2017, Tesla CEO Elon Musk made a bet* that he could help stabilize the region and prevent blackouts by backing up its renewable sources with Tesla Powerpacks, which he then won handily.[9] The farm can power 30,000 homes for up to an hour, has a storage capacity of 100 MW, and has saved Australia millions of dollars. Another prime example of this is on the island of Kauai, Hawaii, where Li-Ion battery storage backs up solar installations. Furthermore, Li-Ion battery storage does have an advantage in that it requires a small amount of land and it is very scalable and repeatable, making construction quick and cheap.

Though these are promising case studies for Li-Ion battery grid storage, the future is far from certain. Li-Ion batteries have serious lim-

---

* The bet was that Tesla could install the required capacity in 100 days from the signing of the contract or it would be free. The contract was completed in just 60 days.

itations. First, these batteries degrade over time. Each cycle (when it charges and discharges) slightly damages the battery and therefore over time it becomes less efficient. This is one of the main reasons why your phone loses charge so much faster after you've had it for a couple years than when you first bought it. This means Li-Ion battery storage will require constant evaluations and repairs to keep it working properly and economically. Additionally, costs are comparatively high for this storage, though the technology can still be greatly improved upon and costs are falling. Environmental impacts from these could become significant if not handled properly, an issue that is already being seen in solar panel waste and is becoming increasingly worrisome from electric vehicles.

Solid-state batteries work in a similar fashion to Li-Ion batteries but with a couple key advantages. While Li-Ion technology uses a gel electrolyte to move electrons around, a solid-state uses solids for the components of the battery. Among other advantages, this results in high efficiency, higher capacity, and operations at a cooler temperature. While still in development, solid-state batteries are a near-future technological solution and could revolutionize electrical battery storage. They could also have massive implications for electric vehicles, which are already becoming competitive to gasoline vehicles (see chapter 13).

Pumped-Hydro. While Li-Ion batteries are gaining increasing attention, it is pumped-hydro storage that currently provides, by far, the most grid backup. The premise of pumped storage is simple. When there is an influx of excess energy, the power is used to pump water up a hill to a reservoir. When energy demand is greater than output (like at night when solar panels are not producing power), the water is released from the reservoir where it flows down the mountain to spin a turbine and then a generator, which will produce power. Once at the bottom, it enters another reservoir where it can then be pumped back up again.

In 2019, there was nearly 160,000 MW of installed pumped-hydro storage capacity around the world.[10] Of this, 43% (~68,000 MW) was just in East Asia and the Pacific regions.[11] Due to the nature of pumped-hydro storage, there is a massive area of land required. Since water is not very energy dense, the reservoirs need to be very large in order to produce enough power. To get the most "bang for your buck" from this technology, the distance between the upper and lower reservoirs needs to be as much as possible—so that water flowing down can gain more energy from gravity. These factors make expansion of pumped hydro difficult. The high capital costs and very specific criteria that need met means there is only a limited number of places where pumped hydro can be utilized. Though there is still limited expansion available for this type of storage, it is not going to work in many regions around the world. Climatic changes could also impact the future applicability of pumped hydro (see chapter 4 for a similar comparison on how climatic changes are impacting hydroelectric power).

Other Storage Solutions. Aside from the previously mentioned storage solutions, there are a multitude of experimental or upcoming storage solutions. Mechanical batteries have various designs, but one of the most promising types is to use excess energy to spin a flywheel at a later time, in turn producing power. Compressed Air Energy Storage (CAES) uses extra power to compress an air, which can be then released to spin a turbine when needed. Gravity batteries use energy to elevate a weight and can release the weight to produce energy when needed (this is the same basic principle that pumped hydro works off, just using solids instead of liquids). Thermal batteries use excess electricity to create heat, which can be used in various ways to produce power. Additionally, liquid-flow and liquid-metal batteries use a similar process to Li-Ion batteries but use a liquid instead of a solid or gel. Clearly, the world is at no shortage for ideas.

Peaker Plants. Peaker plants (here-after referred to as "peakers") are small power plants that are meant specifically to meet extra electric demand that the normal grid can't handle. Because solar and wind are intermittent, they are more reliant on peakers to fill the gaps in power supply than other sources. This drives up costs of production and since these almost exclusively use natural gas, increases carbon emissions. Peakers are a necessity as long as there is not an effective storage system and therefore are simply another reason why solar and wind will not be no-carbon solutions until the storage system issue is solved. Because these plants are only operational during peak demand (hence the name), the electric they produce is significantly costlier than baseload or load-following power production. This is one of the reasons why California's electric prices are so high, along with the closure of its nuclear plants, power outages, and various other factors. Generally, hydroelectric dams and geothermal plants do not require much, if any, peaker plant backing because the power coming from them can be easily ramped up or down according to demand and are very consistent. Some thermal baseload plants—such as nuclear or coal—require peakers, though since they are more consistent the trends of when and where peakers are required are substantially more predictable and are utilized less.

Energy Transmission

If we could say, move the excess energy being produced by California's solar farms to New York in a quick and economical way, everyone would win. The biggest hurdle here is basic physics. As electricity is moved through a transmission cable from the power plant to your house, some of the electricity is lost through resistance heating with the cables. The US Energy Information Agency (EIA) even estimates that about 5% of transmitted electricity in the US is simply lost this way every year.[12] This is not to say that it's impossible to transmit power over long distances, but with current technology it isn't economical or efficient. In a perfect world, every power connection in the country would be connected to an intelligent grid, regulated by an artificial intelligence (AI), or advanced computer algorithms. This AI would be utilizing real time data to distribute power as effectively and quickly as possible but would also consider future weather patterns and predictable factors to determine when and where there may be changes in the energy supply. A system like this is what is often called a "Smart Grid." A smart grid, though extremely expensive, could have immense economic benefits.

If you paid attention during the 2021 Texas blackouts, you may already know that the US has three main interconnected grids. There is an eastern, western, and Texas grid. States on the east grid can transfer power freely between each other and the same with the west grid. Texas's grid is independent and does not allow for an easy transfer of power from other grids, which only further exacerbated the blackouts. These grids are a marvel of engineering, but in the modern day they are becoming outdated. Additionally, the separation of the eastern and western grids does not allow for energy to be easily transferred from California to New York like in our example. A smart grid would further integrate these grids and help regulate them to be more efficient. Of course, it is because of the worry of federal regulation that Texas has its own grid in the first place.

A report issued by the Electric Power Research Institute (EPRI) in 2011 noted that a smart grid for the US could cost $338-476 billion,[13] but would also result in benefits that could save the US $1.3-2 Trillion.[14] A big price tag for a big reward. This smart grid would also potentially reduce greenhouse-gas emissions. Other benefits could also include national security and energy storage improvements. High-voltage, long-distance, and underground transmission cables will also greatly improve the efficiency of power transport over greater distances. For an increasingly intermittent power supply, the further development and installation of these will be imperative for maintaining a stable power grid.

It is worth mentioning that long-distance transmission is not impossible, just incredibly difficult. China has been working on (and utilizing) long distance transmission lines through Ultra-High Voltage (UHV) cables. This has allowed the country to massively improve its energy transfers so that the energy being produced by the hydropower in the west and the coal power in the north can be used in China's large population centers, which are concentrated in the southeast. A coupling of techniques such as those seen in China's UHV systems with advanced battery storage will ultimately allow for a massive buildup of renewable power in the upcoming decades, but is a massive hinderance to their implementation today.

Summary on Introduction to Renewable Power

A smart grid, with the use of other methods of storing renewable energy, and the general improvement of renewable power technology could see a future where it is possible to run a renewable world. However, because of the immense changes and technological advancements that require the practical implementation of this, wind and solar might not be able to reap the benefits of this for a long time to come. In the case of a smart grid, it would be a great idea and a true engineering marvel, but to even get a plan like this passed through the legislative system in the US could take many years. For these reasons it seems that wind and solar have great potential, but many regions around the world will be unable to fully tap into this until much later in time. Therefore, wind and solar are currently most applicable for use in rural areas around the world—likely alongside with one another*—due to the low power demands and plethora of space. These can be great when applied properly, but it is highly improbable that they will be able to fill the shoes of fossil fuels, nuclear power, and hydro for more densely populated areas in the near future. Once an effective and widely deployable storage system is achieved, the expansion of wind and solar could very well become the cheapest form of energy production.

This is not a reason to say that wind and solar have no place. Because they certainly do. Even when things such as the toxicity of materials used in solar panels and how dangerous and bad for the environment mining for things such as cobalt, tin, sand, and other resources used in these types of energies is accounted for, wind and solar are far cleaner than the fossil fuels being used today and are becoming increasingly cost-competitive.** The main issue is that because of the energy storage dilemma, these can become extremely expensive projects when used outside of their proper context. Theoretically, the technology and resources exist to power the world off wind and solar. However, this raises a key point: just because we can, doesn't mean we should.

The global energy transition is going to be the largest united project the human species has ever tackled. So, it will be essential for governments around the globe to not just do what's required, but to do it in a way that is positive for the economy and society. And although solar and wind are great in their own ways, forcing them to be used outside of areas that they excel in could result in massive economic and environmental losses in many countries around the world.

* Wind and solar work well together in many cases since winds tend to be strongest at dusk and dawn while solar is obviously strongest between these points.
** Because renewables are often overly glorified, these next several chapters will focus on the downsides. However, it is imperative to keep this fact in the back of your mind.

Chapter 2

When The Sun Shines
Solar Power, a Modern and Future Solution

In 1964, Nikolai Kardashev, a soviet astrophysicist, proposed something called the Kardashev Scale. The Kardashev Scale is a hypothetical scale that attempts to measure how advanced a civilization is by assessing its total energy production. The reasoning behind this suggests that as we become increasingly more advanced, we will use an exponential amount of energy as a civilization. That being said, Kardashev classified his scale onto 3 levels of society ranking from lowest (a Type I civilization) to highest (a Type III civilization). According to this scale, just for a civilization to become a Type I, they must be able to use and store all of the energy available on its host planet from its parent star.[1] We are very, very far away from becoming even a Type I civilization (also described as a planetary civilization). Considering that the Sun delivers more energy to Earth in under 2 hours than we use to power our planet for a year, we are, at the very minimum, centuries from being anywhere close to being capable of achieving this—and likely millennium away.(See note [2]) But hypothetically, if we were to do this, we would have complete control over the Earth. It is also not far-fetched to say we would also have some control over other planets and forces in our solar belt at this point in the future. Going through these "levels of civilization" may very well end up being humanities destiny.

In order to do this, there are only two distinct forms of energy generation that could support a civilization of this scale—nuclear fusion, and of course, solar power. Solar power could very well be the perfect energy source for a space faring civilization of the future, and may end up playing an integral part of our being as we advance into new levels of civilization. If we were to even think about progressing to a Type II (Stellar Civilization) or Type III (Galactic Civilization), we would need an immense advancement in energy harnessing technology, but hypothetically, it is not impossible.

Along with this futuristic theme, solar power could be the technology used for one of the largest theoretical contraptions ever conceived—a Dyson Sphere. A Dyson Sphere is a megastructure that would surround a star and be capable of capturing all or most of its energy. An advanced form of solar power like this could be the ultimate form of power for a highly advanced civilization.

Environmental activists repeatedly claim that solar power could be the perfect energy solution for our entire civilization. Along with wind power and other renewable energies, solar panels are simply worshiped for being the perfect energy source. It's 100% clean energy; it has little to no impact on the environments they are built in, and there are obviously no hazardous emissions that come from them—not to mention how affordable they are becoming compared to other energy options. The sad truth is, that although solar energy may be a perfect energy source in the future, it isn't there yet and won't be for a long time to come. Regardless of the exponential progress being made in the solar industry, solar energy is unable to save us from climate change by itself.

Solar Power Technologies

There are two distinct ways of producing solar energy. The first is photovoltaic (PV) solar power. This is the most common and most popular form of solar power and is generally what you imagine when you think of "solar panels." In simple terms, PV solar takes in sunlight and converts it directly into an electric current that can be harnessed. The second type is concentrated solar. This essentially uses mirrors to reflect sunlight at a central location where the focused heat can be used to for a steam turbine. This is seen most notably at large solar farms where a lot of heat can be concentrated at a single point, such as at the Ivanpah solar farm near Las Vegas. Thermal solar power is not very commonly used in civilian residential applications, though it is sometimes used for heating. Throughout this chapter, PV solar will be mainly discussed, but distinctions will be made when necessary.

When you look at the data, several problems with solar technology become quickly prevalent. Obviously, solar panels only produce electricity when the sun is shining (in some areas this can be as low as 10% of the time). This creates many obvious problems and more not so obvious ones. The most obvious issue is how do we produce power when the sun isn't shining? There seems to be two main schools of thought on this that are distinctly different but also will require overlap for effective use.

First, we just add more solar energy, especially into areas where they will fit in with minimal disruption—like with rooftop solar panels—and send excess energy to other regions. The problem here is that it is not always realistic to do on a large industrial scale. Yes, you could cover every roof in the world with solar panels and that would help people save some energy and money—especially in the summer months—but only 16% of total energy consumption in the US in 2018 was non-urban residential (essentially meaning suburbs).[3] This is the ideal place for retail solar since there is so much rooftop potential. Indeed, 16% is a fantastic starting point, but the big changes would need to occur within cities.

Large swaths of the population in the US and other developed countries live inside of big cities. In 2013, the world's urban areas accounted for about 64% of global primary energy use and produced 70% of the planet's carbon-dioxide emissions.[4] The problem here is that rooftop solar panels often cannot cancel out the energy consumption for the one household they are providing for. So, to try to cover a housing complex with enough to even make a dent in the total energy consumption is not realistic. You could cover every building rooftop in the world with solar panels, but this would still not come close to solving our energy requirements. This would produce an enormous amount of energy, but relative to the expense and time something like this would cost, it isn't realistic in a manufacturing or economic sense (especially on short timescales). The best path of implementation for this would be to build new houses with solar panels already installed, while also providing incentives for people living in sunny regions to get solar panels on their roof.

More to the point, in a system where huge amounts of excess utility scale solar are being produced and transmitted between regions, there remains the current problems of figuring out a way to transmit this power cheaply, quickly, efficiently, and consistently while also being wary of the massive amounts of land that is required.

Solar Energy Storage Issues

The second solution to the inconsistency of solar power is to store excess energy not used while solar panels are producing energy. The problem here is that battery technology is extremely far away from being able to hold the enormous amounts of energy required (see chapter 1). In fact, when the sun is shining, solar farms often don't even come close to storing all of the energy they receive. A prime example of this is in California. California is regarded as the United States leader in clean energy initiative and has invested extensively into solar power. Now, because they are not able to store all the excess energy being produced during sunny periods, they've run into a brick wall of problems that were not anticipated. Because of the inability to store large amounts of excess energy in batteries, whenever it is an especially hot and sunny day, California can actually encounter power surges coming from the solar farms that have been built. If left untamed, these power surges can—and will—destroy electric grids across parts of the state, causing massive blackouts on a regular basis. To prevent this, California frequently pays neighboring states such as Arizona and Nevada to take the excess energy away. As the *Los Angeles Times* reported:

> "On 14 days during March [2017], Arizona utilities got a gift from California: free solar power. Well, actually better than free. California produced so much solar power on those days that it paid Arizona to take excess electricity its residents weren't using to avoid overloading its own power lines. It happened on eight days in January and nine in February as well. All told, those transactions helped save Arizona electricity customers millions of dollars this year, though grid operators declined to say exactly how much. And California also has paid other states to take power. The number of days that California dumped its unused solar electricity would have been even higher if the state hadn't ordered some solar plants to reduce production — even as natural gas power plants, which contribute to greenhouse gas emissions, continued generating electricity. Solar and wind power production was curtailed a relatively small amount — about 3% in the first quarter of 2017 — but that's more than double the same period

last year. And the surge in solar power could push the number even higher in the future." [5]

While it should be acknowledged that fossil fuel and nuclear plants are in many cases being shut down due to the increase of energy coming from renewables, it is clear that the addition of renewables without concrete storage plans doesn't allow the state to consider a future that is 100% rid of fossil fuels—in particularly natural gas. It is imperative to point out that California has to actually pay the neighboring states to take the energy it created, instead of the states paying California for said energy. In short, the inconsistency is costing California in a big way. While not all of this is strictly due to solar power technology (much of it is related to California's renewable energy policies), the current situation remains.

Despite this exportation of power, California is simultaneously by far the largest *importer* of energy in the country—in 2019 importing about twice as much energy as the second biggest importer: Virginia.[6] This is because they have shut down so many power plants to make room for solar that when the sun isn't shining, they have to import large amounts of energy from other states, which again, costs money. These imports are almost entirely made of coal and natural gas produced electricity. This means that although California is often praised for making a real movement towards clean energy, they are, in fact, far from being rid of fossil fuels and have simply put much of their carbon-based energy onto other states.

As discussed previously, it isn't practical to transport the electric in California all over the country when there is excess. Due to thermodynamics, when electricity is transported through power lines, it loses some of its power through heat. This is part of the reason why we have power plants stationed intermittently throughout the country. Fixing this would at least allow for California to export electricity at a profit to other states.

Largely (but not only) due to the massive expansion of renewables (particularly through solar farms), retail electricity rates in California in 2019 were an astounding 60% higher than the national average.[7] Comparing California's electricity prices to other states in the region, they are astronomical. Electricity rates in California were 92% higher than Nevada, 91% higher than in Oregon, 60% greater than Arizona, and over twice as much as retail prices in Washington state.[8] Granted, California is not home to the country's highest electric price rates. Alaska, Connecticut, Hawaii, Massachusetts, New Hampshire, and Rhode Island all had costlier retail electric than California. There are a few reasons to explain this. Alaska and Hawaii are obviously disconnected from the 48 continuous states, which massively increases the cost. The other states mentioned are all concentrated in the New England area, and costs are high due to their reliance on the importation of power (such as hydroelectric from Canada). Accounting for the average of all New England states, California has the same average price of electricity as this region of the country for much of the year.[9] To be fair, other things such as damage to electrical infrastructure from wildfires and regional inflation contribute to California's prices.

It's not that California prices have always been this much higher than the national average, either. In 2012, just at the start of the states' massive solar expansion, its retail prices were 1.37 times that of the national average. By 2016 this had risen to 1.46, and was an astounding 1.60 times higher than the national average in 2019.[10]

But it isn't just the Golden State that has these problems. Germany is one of the world's leading countries as far as renewable energy infrastructure and governmental legislation. However, Germany experiences many of these same problems. In fact, in 2019 about 34% of Germany's electricity was being produced by renewables (9% being solar)[11] and they paid some of the highest electricity prices in the developed world. From 2006 to 2018, Germany's electricity prices rose by 51%.[12] Even as Germany strives towards its goal to produce 80% of its electricity with renewables by 2050, it could end up being hindered economically in the very real possibility that solar power issues aren't fixed. (To be fair, while it is true that solar has proven to be very costly in many regions, it is also important to note that one explanation for this cost can be that early adopters have paid massive amounts into funding for the development of the technology.)

Because of this unreliability and due to the fact that wind and sunlight are not very energy dense (when compared to coal, uranium, natural gas, oil, or even water), the more we depend on solar and wind, the less cost effective they become because of the need for baseload power supplies. Solar is fantastic at helping to provide energy during fluctuating times of day, but it is terrible as a baseload power source. It becomes exponentially more difficult and therefore expensive as we try to power these utilities during times when solar production is not at its peak. The Duck Curve is a good visual example of this relationship. The Duck Curve basically depicts that the biggest issue for solar power is that electric demand is typically at its highest in the evening, which is unfortunately right when solar production is decreasing. Solar power production is at its highest around noon, when it is not needed as much, causing it to be curtailed for reasons previously discussed. (In extreme cases, this Duck Curve can transform into something else coined as the Nessie Curve, which applies the same theory but is even more exaggerated.)

Although solar is becoming cheaper every year on a panel-per-panel basis, if we increase our dependence on it, it may become more expensive at exponential rates. The opposite of most technologies, solar actually becomes more expensive for our power grids the more of it we have, even as it becomes cheaper to manufacture, construct, and install. This issue is still apparent but much less drastic with regards to wind power.

Environmental Issues of Solar Power

Tying in with all of these energy production problems creates another, not so obvious problem. Solar farms create a major disruption to the ecosystems that they are built in. Often, they are built in wide open fields or deserts so one would expect this would cause minimal disturbance as opposed to things like cutting down the Amazon, but deserts and fields also have important ecosystems that need protected, many species in which are often endangered. Especially when considering the sheer amount of land consumed in solar farms, if we continue to further expand these energies, we will cover so much of the Earth's surface that we could start to have serious ecological problems. When these solar farms are developed, the natural ecosystem has to be relocated to somewhere else—often into captivity and resulting in a large fatality rate among disturbed species. For example, an anonymous scientist who aided in the development of the Ivanpah solar farm (the largest solar farm in California, producing about 10% of the states electricity) reportedly said, "Everybody knows that translocation [of desert tortoises] doesn't work. When you're walking in front of a bulldozer, crying, and moving animals and cacti out of the way, it's hard to think that the project is a good idea." [13]

There is also the problem with what happens to a solar panel after it breaks. PV solar panels contain heavy metals that can be very toxic (such as lead and cadmium). With the absence of a proper recycling plan in most countries, these toxic wastes often end up in third world countries, being taken apart by hand and poisoning the local populations. A study by Environmental Progress in 2017 found that, per unit of energy, solar panels create three-hundred times more toxic waste than nuclear energy[14] and the International Renewable Energy Agency (IRENA) estimates that at the end of 2016, there were 250,000 metric tons of solar panel waste, and that there could be 78 million metric tons of this waste by 2050.[15] Furthermore, because of the toxic materials, this waste cannot (or more accurately, should not) be put into landfills and thus often results in the issue of the materials being exported and poisoning communities in other countries. In the cases where this waste is not exported, it can become a serious groundwater contaminant in landfills. In any case, this is to say nothing of the massive amounts of rare earth metals that need mined for solar power (and many other electronics for our modern world) which have notorious ecological and humanitarian downsides.

Despite this, the solar waste problem still has an easy solution—an effective recycling program. While recycling programs are not yet very common, they could remove significant amounts of problematic waste and make the manufacturing of panels cheaper and more sustainable. It should also be noted that many of the large solar farms actually do not have this problem on such a large scale because they utilize thermal solar power, where most of the waste is just glass.

Improving Solar Power

Solar technology is still improving; it is making great strides in every aspect of the industry and every year solar panels become cheaper and more efficient, but this isn't enough. This is simply because in order to beat climate change with renewables we are going to need to be capable of creating these renewables with energy from renewables. This is a major obstacle that is constantly overlooked. Because of the lack of energy reliability produced by renewables, it seems highly unlikely that we will be able to produce them in a clean manner, unless produced by a more consistent form of clean power like nuclear or hydroelectric. Although the solar panels will cancel out the harmful emissions created during manufacturing over their lifetime of zero-carbon power generation, it makes the objective much harder if we are still always emitting pollutants into the air when they are manufactured (although obviously solar power still has a much lower carbon emission over its lifetime than many other forms of energy).

There is, however, light at the end of the tunnel for solar power. It is already a massive industry and technological progress is being made constantly. There have been hypotheses and proven experiments on various ways to make solar power much more efficient and applicable, which we will take a look at now.

Improved Technology. The use of carbon nanotubes in particular has shown to greatly increase the efficiency of solar panels. Currently, a typical rooftop solar panel only turns approximately 15-18% of the light that hits it into electricity, but most panels on solar farms have an efficiency between 20-25%, and most of the rest of the light escapes the panel in the form of heat. Through the use of carbon nanotubes inserted into the panel, much of this lost energy can be acquired again and used for energy; this could possibly take solar panels up to a theoretical efficiency of 80%. If this were achieved, it could indeed become a game changer for the solar industry.[16] Other higher grade materials could be used for solar panels (as are the ones attached to International Space Station and other satellites) to increase efficiency, but the costs here are prohibitive and not widely applicable for commercial use. Another method that is gaining slight traction could be to develop solar-panel type devices that absorb a wider spectrum of light.

Beam Focusing. One other experimental tech for solar energy is the process of focusing sunlight into stronger strands. Similar to light intensifying under a magnifying glass, this method looks at how applicable it may be to focus light that would otherwise be lost into a strong beam to be more effectively used by solar panels. It has been proposed that effective use of this technology could make solar panels more efficient by a great factor. Through this reasoning then, we may need much less space to produce equal amounts of electricity and the cost of solar panels would be dramatically reduced. Considering that solar farms use many magnitudes of more space than traditional power plants, the lower use of space will greatly benefit this industry.

Space-Based Solar. Potentially putting solar panels in space and then beaming down the energy in the form of microwaves could prove very useful. Large amounts of global energy demand could be met by this method, but there are concerns over how much it would cost to put all of this material in space and also over what to do once the panels are not efficient anymore (after 20-30 years on Earth most solar panels need replaced) and possible space debris caused. Overall, space-based solar could be a game-changer (especially due to the rapidly decreasing costs of space-travel) but is a long-term future technology that will not be able to help humanity with the current climate crisis very soon. There are also concerns about large amounts of space debris caused by this, which raises alarms about a potential runaway debris cloud in space, commonly known as the "Kessler Syndrome," which could destroy many or all of our satellites and render space travel impossible for several generations.

Solar Shingles. The idea of creating photovoltaic roof shingles has been gaining significant traction. Instead of having a roof and installing PV solar panels overtop, what if your actual roof could be PV? Tesla and a couple other companies are currently working on this. It could potentially be cheaper, easier to install, and will look *significantly* more aesthetically pleasing. It is plausible that solar shingles will begin to dom-inate in residential housing in the next couple decades while traditional PV panels will start to focus on utility-scale power generation and small scale businesses such as on farms and in other rural areas.

Transparent Solar Panels. Transparent solar panels could greatly increase the potential for solar energy harvest while also minimally interrupting existing infrastructure and land usage. The idea of transparent PV solar has been around for some time now, and it is starting to make progress. This technology would allow for say, windows that also generate power. This could be massively beneficial to dense urban areas where there is a plentitude of available window space on the side of skyscrapers that could be used to produce power. If this technology is ever commercialized, it will not be for several decades.

Summary of Solar Power

Unfortunately, while these and many other technologies are very exciting, they are still very much in their infancy. There is a great deal of research needed to be put into them still, but if we could get these to work, and especially if we could get them to work cooperatively (which is far from guaranteed), then we could create solar power that is exponentially more effective and applicable than modern day panels. These would take away some of the problems that the solar industry faces today and into the future. Today, the biggest hurdles for solar energy remain as being the sun only shines so often, and we don't yet have an effective energy storage system.

Despite the fact that most of this technology will likely mature, there is a very real chance that it might not be in time. We need immediate solutions to climate change, and solar is almost certainly behind the curve to arrest our energy system single handedly. Even if we invest in advanced solutions right now, then there is a very real chance we won't be able to catch up to where we need to be before the climate is completely destroyed. Until solar technology is far better, it seems likely that the best option is to use solar wherever applicable and judge its limits appropriately, but to not get caught up in the craze—accepting that we need other solutions in conjunction with this. To continue attempting to power the world from an energy source that can't take the load is the worst option that we have. Currently, solar power is great for powering small towns and helping consumers reduce their electric bills, but it doesn't seem to be the sole solution to fix the immediate problem of climate change at this moment.

In spite of the challenges facing solar technology, solar PV is one of the few technologies under the IEA's Sustainable Development Scenario (SDS, which will be referenced heavily throughout this book) that is on track to meet the emissions reductions goals set for it needed worldwide. Just in 2017, China added a staggering 53.1 GW of PV solar capacity to its electric grid.[17]

Chapter 3

When The Wind Blows
Wind Power, a Modern Solution

Wind power is another very popular form of renewable energy. Although wind and solar energies are often associated together, they are very different and have different impacts, especially under consideration of economics and created waste. Compared to solar power, wind power seems to have fewer obstacles and what problems there are appear to be more solvable in the near-term. Basically, wind power is made from giant blades that are spun by wind currents, therefore spinning a generator, and creating electricity. Wind turbines are placed in areas where the wind blows often and strong. These are typically in open plains and on top of mountains, as well as off the coast in the ocean, referred to as onshore and offshore, respectively. And although solar power makes the headlines, it is wind energy that is making real progress around the world.*

There are a few countries around the world that are taking wind power and running with it. Most notable of these are China, the UK, the Netherlands, Denmark, Germany, the US, and India. For example, in 2018 Denmark produced 13,900 Gigawatt Hours (GWh) from just wind power, making up approximately 46% of the country's electric generation.[1] In 2019, China alone added approximately 20.6 GW of wind power capacity, with 2.3 GW being offshore[2] and added nearly 25% more wind power during Q4 than the US did over the entire year.[3] The US is not slacking though. During Q2 of 2020, the US had 25.3 GW under construction and a further 18.1 in advanced development stages.[4] Globally, in 2019 there was 54.3 GW of added onshore wind capacity and an additional 6.1 GW came from offshore.[5] For

* One possible reason for this discrepancy is because consumers now have the option for solar power on their homes roof, which is not the case with wind power.

comparison, the US had around 1,100 GW of installed capacity (between all energy sources) at the end of 2019.[6]

As 9% of US electric capacity is nuclear, the world added the equivalent of nearly two-thirds of the total nuclear capacity in the US through wind in a single year. This should not be surprising either, as wind power is becoming increasingly more cost competitive due to economies of scale and improving technology. Wind benefits not simply from the manufacturability that also helps solar, but also from the fact that it is more consistent than compared to how often the sun is shining (especially offshore winds). This makes wind power much more capable of providing consistent power than solar. However, as with any form of energy production, there are some downsides.

Environmental Impacts of Wind Power

There are various environmental concerns about wind power that are worth discussing. However, these downsides are often blown out of proportion by the opponents of wind power. To fairly assess the potential of wind power, these issues need addressed upfront and foremost.

Threats to Bird and Bat Populations. One of the biggest arguments against wind power is that they are very harmful to bird and bat populations, and wind turbines have been frequently observed to kill many flying animals. Although these farms do kill many birds and bats, the amount that they kill is insignificant when compared to the number of birds that fly into windows or are even killed by house cats every year.[7] The part of this issue that is often avoided, however, is that much of the time it is the larger predatory birds (such as eagles and falcons) that are affected the most. This where the real problem is. Many large birds of prey already have dwindling populations, and the use of wind turbines have affected them further. Considering how large of an impact that birds of prey have on their own ecosystems, it is an actual issue that needs addressing.

Unlike many problems with other energies, this has begun to be addressed and for the most part, is being properly handled by both the scientific and engineering communities. Before significant research and observation was put into this issue, wind farms were indiscriminately put wherever the wind blows the most. Now, wind companies are starting to give notice to bird migration and hunting patterns. This is expected to help reduce bird and bat deaths significantly in the future. One could hope that this would go forward to include the threats that face large birds of prey, but it is hard to say with certainty.

For example, "In December, 2016, a new eagle-management plan announced a final rule by the federal government that would give wind energy developers 30-year permits to 'take' or incidentally kill protected Bald and Golden Eagles, without requiring the industry to share mortality data with the public or take into consideration such critical factors as proper siting." [8] This could obviously show to be a problem if wind power companies decide to not consider the problems as a serious concern.

Various solutions other than turbine placement and policy are in the works as well. This includes painting one of the blades black instead of white—which helps birds to notice the spinning of the blades and avoid them—and installing machine learning algorithms and cameras on the turbines that makes the blades slow down when they detect an animal flying near them.[9] Interestingly, another problem that is beginning to show is how much of an impact wind turbines can have on insect populations. Many insects use wind currents for migrating, and since these turbines are set up where wind currents are strong, there have been observations of significant decline in insect populations due to this. Furthermore, the killed bugs on the blades make them less aerodynamic and they therefore then produce less electricity. This issue could impact the wind industry in the future as a clear solution is yet to be identified.

Offshore wind farms are also a large emerging part of the industry but show similar effects on bird populations as regular wind farms. They seem to have minimal effect on fish life as well as sea-fairing birds since there are not as many endangered marine birds and these species hunt over greater distances. There is even actually some slight evidence that offshore turbines can help sea life populations by acting as a sort of artificial reef.[10] Of course, there may be some circumstances where offshore wind is harmful to the ecosystem, but generally speaking it is the same or even better than onshore impacts.

Space Occupancy. Like solar power, wind takes up huge amounts of space. Wind turbines themselves are massive by nature, and these wind farms can occupy very large areas. This is only a problem if we look at it in the wrong light. Although they take up a great deal of space, that doesn't mean that the area is unusable for other things, as is often seen with solar farms. In between each turbine there must be a space of 5 to 10 rotor diameters. (This is so that wind disturbances by each turbine is minimized. If they are too close to each other they become less efficient, and if they are too far, they require more space. The distance between each turbine varies widely on a site-to-site basis.) Therefore, there is a lot of unused space between each turbine. In truth, wind turbines only actively consume a fraction of the space that the overall farm occupies as a result of this spacing.[11] Because of this, there is minimal impact of the space occupancy of these farms, and it is completely possible to produce agriculture, areas for livestock, and other small-scale infrastructure inside of these farms. Since wind turbines are so large and only produce power while the wind is blowing, they are often built in the countryside where there are large open plains. This makes them even more perfect for powering small towns, farms, and other industries that fit well with what the space could be potentially used for.

Greenhouse-Gas Emissions. Like other renewables, wind power has a very small carbon footprint. On average, wind energy emits an average of 13 grams of CO_2 per kilowatt hour (g/kWh) at the low end,[12] but some other estimates place it as high as over 34 g/ kWh.[13] Coal on the other hand produces, on average, 909 g/kWh.[14] Dividing these numbers, coal power plants produce nearly 70 times more CO_2/kWh than wind power. Obviously, this is due to the fact that, like solar, wind creates little to no emissions while operational. The CO_2 that is produced from wind turbines is nearly all created during manufacturing and construction, because of the massive amounts of steel and concrete required for them. Additionally, offshore wind farms typically produce more CO_2/kWh because much more material is required for construction in the ocean. Like with solar, to go completely carbon neutral wind turbines need to be created by no-carbon technology. This again is going to be quite a hurdle for low-carbon energy in the future. Although wind power is showing to have a much smaller carbon footprint than many other energy solutions, there is a threat to this claim that has been gaining some attention. This problem is a substance called sulfur-hexafluoride (SF_6). SF_6 is a gas that isn't flammable, toxic, or corrosive—giving it many applications. It is also a very heavy gas.**

These qualities make SF_6 an extremely good electrical insulator. In electrical enclosures, power surges and other anomalies can cause electrical arcs to shoot off inside of the containment, damaging or destroying the product. The use of SF_6 with electronics in these scenarios reduces the likelihood and damaging effects of electrical arcs.[15] Before SF_6 was widely used, these arcs would often destroy the equipment, sometimes causing fires to break out or even explode.

However, these properties also make SF_6 one of the most potent greenhouse gases that we know of. SF_6 is such a strong greenhouse gas that it has a warming potential 23,000 times greater than CO_2 and can stay in the atmosphere for over a thousand years.[16] This gas is typically well contained and even recycled, but when it leaks, the effects can potentially be very bad. This gas is used in parts of wind turbines, and that is the reason why there has been an uproar from some against wind power. These claims however are severely un-justified. Yes, wind power does utilize this gas and yes sometimes it can leak, but the bigger problem is not the wind industry, but rather the electrical industry as a whole. SF_6 is used in a large variety of electrical equipment and wind power is a very small aspect of this market. Additionally, not much of it is used and even less actually leaks from containment.

In 2019, the Wind Europe association responded to a BBC article from the same year[17] discussing this topic. While acknowledging that wind turbines can leak SF_6, Wind Europe also made sure to counter argue by stating that "Data from Vattenfall suggests leakage emissions from Europe's 100,000 wind turbines were about 900kg of SF_6 over the last six years. This is equivalent to 3,525 [metric tons] of CO_2 a year. …By comparison wind energy avoids the emission of 255 million [metric tons] of CO_2 in Europe a year by generating 336TWh of electricity displacing fossil fuels. The SF6 leakage therefore represents around 0.001% of the emissions avoided thanks to wind energy every

---

* Like how ingesting helium (a very light gas) can make your voice very high pitched, SF_6 can make your voice very deep.

year."[18] Furthermore, there is also extensive research occurring to replace SF_6 and several substances are showing promise. As wind turbines only make up a small amount of the use of SF_6, the effects should not be tied directly to them.

Offshore Wind

There is a fair amount to be excited about with offshore wind. Bigger turbines can be built for offshore purposes, allowing for more energy production, and there are large swaths of ocean that are not occupied—like they are on land—with more consistent and stronger winds. With recent developments in wind technology and marketing, offshore wind is poised to boom over the next couple decades. While offshore wind is behind the SDS curve, it is possible for it to catch up or at least make off far better than many other technologies will. In the SDS, offshore wind generates 606 TWh in 2030, a substantial increase from today.[19] As is the case with most renewable technologies, the potential energy that could be harnessed from offshore wind is immense. Globally, offshore wind could theoretically provide 18 times the current global electricity demand.[20]

Europe is leading on offshore wind capacity. In 2018, the EU had 19 GW of installed capacity, while China had just 4 GW and the rest of the world had essentially none.[21] Despite this, the technology is well positioned to become a major renewable energy provider in coming years. "Offshore wind is in a category of its own, as the only variable baseload power generation technology. New offshore wind projects have capacity factors of 40%-50%, as larger turbines and other technology improvements are helping to make the most of available wind resources. At these levels, offshore wind matches the capacity factors of efficient gas-fired power plants, coal-fired power plants in some regions, exceeds those of onshore wind and is about double those of solar PV."[22] (Capacity Factor is essentially a way to measure how often something is generating power. 100% means it was producing energy 24/7 over the entire year.)

Because these turbines are in the open ocean, construction can be difficult, but this is a process that will only continue to improve upon itself, with one study finding that installation times of the turbines and foundation dropped by 70% between 2000 and 2017.[23] We are also starting to see new designs for offshore wind where the turbines float in the ocean (very much like oil rigs today), which will make their implementation easier due to them not requiring a foundation.

Summary of Wind Power

Wind power has massive potential. The advantages of manufacturability, higher consistency than solar, and large potential for expansion gives wind a good heading. Surprisingly, in 2016 wind power surpassed hydroelectric to become the renewable power with the most installed capacity in the United States.[24] However, wind still needs backup storage power just like solar.

In the case of a severe storm, wind turbines have to shut down. This cut-off point is typically around 55 mph.[25] As storms will get stronger due to climate change, this could become a major issue in stormy regions. The fact that steel is a major component of wind turbines is encouraging because this makes them highly recyclable. (Steel is one of the most recycled materials in the world. It can be recycled indefinitely with little-to-no loss of quality.) Because of this, waste from wind turbines could build up, but large amounts of this are relatively easy to recycle and are non-toxic.

The deployment of wind at large scale is beginning now. Wind is one of the few technologies discussed throughout this book that both has the potential to massively revolutionize our energy mix and is already developed enough for large-scale implementation. There are a plethora of areas where wind is not feasible, but no energy solution is perfect. It cannot completely decarbonize the world by itself, but it can provide a massive help. (As mentioned previously, global potential for wind is 18 times that of current energy demand, but the vast majority of this cannot yet be feasibly utilized.)

There is also a growing interest in small-scale wind power (a similar trend that can be seen within the hydropower industry, as is discussed next chapter) which could greatly increase opportunities of practical energy capture and allow for the average person to utilize wind power for powering their homes. Like with solar power, this would create a market for both small projects (used to produce power for a single home) and large-commercial projects (used to power an entire city). Ideally, this competition will drive prices down while also increasing the performance of the technology. Small wind has an added advantage over solar because of its higher capacity factor (the wind still blows at night). Because of winds higher capacity factor and different peak production hours as solar, the Duck Curve explained in the previous chapter has a smaller impact on it than with solar.

Chapter 4

Where The Water Runs
Hydropower, a Modern & Near-Future Solution

Hydroelectric power is historically the most widely used form of renewable energy and has played an integral role in power production before we even first started using power plants to produce electricity. Today, hydroelectric dams provide a large portion of the electricity for the world and remain the most prominent form of renewable energy—although wind power is looking to take the crown. In 2019, 38% of the utility-scale electricity provided by renewables in the US was from hydropower alone.[1] Other countries like China, Brazil, and Canada use even more hydro-based energy than the US. In 2012, China produced almost three times as much electricity with hydroelectric power than the US and continues to lead the world by far.[2] Brazil has also achieved quite a feat: In 2017, Brazil produced nearly 71% of its total electricity through hydropower, with an additional 21% coming from biofuels, nuclear or wind power.[3] It seems that Brazil has almost become carbon neutral in terms of electric generation without experiencing many of the problems being seen in Germany and California, and it is almost entirely because of the country's vast use of hydropower.

Hydroelectric dams are one of the first historical example of true renewable energies being utilized. By simply taking the forces of a river to turn a generator (much in the same fashion that wind turbines utilize with the air), hydropower is able to effectively turn up to 90% of the energy from the river into usable power,[4] making it far more efficient than solar or wind. Like solar and wind, hydropower takes advantage of the Earth's natural forces to provide energy and therefore does not produce any direct GHG emissions. Because of this, anywhere that has large amounts of running water can take advantage of this system. Technologically, hydropower has not seen much innovation over the past 50 years, and most innovations in this field have recently been focused on safety and construction instead of the inherent technology—

although there are now initiatives working on expanding the possible areas where hydropower could be used, as is discussed later in this chapter.

Aside from generating electricity, dams have a plethora of other benefits, including water storage, flood control, and providing water for irrigation. In the case of a flood, a dam is able to block much of the effect by holding back the water and can then release the water in a stable manner, so the surrounding area is not flooded. Floods can obviously destroy large areas, and dams regularly help to reduce the impact on towns and farmland from this risk. Even the Romans and Greeks used dams for this exact purpose, and the Chinese have been using dams for thousands of years. Additionally, 10% of American cropland is irrigated with water from dams.[5] Reservoir water is even used for industrial and municipal uses.

However, hydropower has its own dirty secret. One of the largest but least-known problems with hydroelectric dams is that they can be very deadly. This is something that is greatly overlooked. It's strange because we see things like nuclear power and coal as dangerous and unpredictable, but when it comes to the numbers, hydroelectric power is historically far more dangerous than most electricity sources, (though fossil fuels—particularly coal and oil—are still more dangerous, as we will see in chapter 18) and certainly holds the record for the single worst disaster in power generation history, and potentially the worst in human history.

Hydropower Disasters

In the US, there have not been many major disasters related to dams so this isn't something Americans tend to consider, but it is something that must be taken seriously:

> "Hydro power generation has a record of few but very major events causing thousands of deaths. In 1975 when the Banqiao, Shimantan & other dams collapsed in Henan, China, at least 30,000 people were killed immediately and some 230,000 overall, with 18 [GW] lost. In 1979 and 1980 in India some 3500 were killed by two hydroelectric dam failures, and in 2009 in Russia 75 were killed by a hydro power

plant turbine disintegration. Early in 2017 nearly 200,000 people were evacuated due to the potential failure of the Oroville Dam in California." [6]

The Banqiao and Shimantan dam failures were some of the worst man-made disasters in history. These simultaneous dam failures killed more than the atomic bombings of Hiroshima and Nagasaki combined. This was due to poor engineering and plain bad luck (an unprecedented amount of rainfall is what initiated the failure), and was largely unique to the Chinese communist party in power at the time.

Incidents like this make hydropower one the least safe forms of widely used energy, and certainly the most dangerous type of low-carbon technology. Hydroelectric dams are still a very common form of energy production and these disasters, though extensively damaging and horrific, are few and far between. It is important to note that as with all things, we have learned from our mistakes and improve every time a disaster occurs. It does raise the question, however, if there has been such a horrific history with these why has there not been a similar backlash by the public when compared to something like nuclear or an overall negative opinion like with coal? The answer is quite simple. In truth, there has not been much worry because an incident of size like those previously mentioned has not happened in the United States. Whereas the nuclear industry had the incident at TMI, and the coal industry has a long-standing reputation for being dangerous, there has not been a significant event related to the hydropower industry inside the United States.

If the Hoover Dam collapsed tomorrow killing thousands and leaving Las Vegas without a water supply, there would likely be major social movements moving for the reduction and/or removal of hydro-power. But as there has never been an incident to this scale in the US, there is quite little public concern about these issues. Although it is obviously great that the US has never had to experience a crisis like this, the safe history creates an illusion that it can't happen in the future. To be honest though, a dam failure like in China in 1975 is basically impossible in the developed world, because of protocols and ethics. You can draw a very close correlation to the relationships of the Banqiao Dam disaster and Maoist China to Chernobyl and the Soviet Union.

Both were the result of brutal communist dictatorships that did not implement proper safety measures, and neither were because of intrinsic problem with the technology.

Other Concerns

Availability. The largest practical problem facing further expansion of hydroelectric power is its availability. Hydropower requires a strong current to produce electricity. A faster flow of water is better as it can turn the turbines faster to produce more energy. This is also why you will see most hydropower stations are set up at sharp drop-offs (against cliffs, steep hills, and waterfalls) so that gravity can create a stronger flow through them. While this means that they are a very good source of energy for areas where there is a large flow of water, areas that are dryer don't have these same opportunities. Because of the effects of global warming and climate change, many rivers around the world are experiencing reduced water flow and, in some cases, completely drying up. A good example of this can be seen at Lake Chad in Africa. Once one of the largest lakes on the planet, Lake Chad has severely dried up over the past century due to both climatic and indigenous factors. The Lake Chad basin once may have covered as much as 8% of the African continent, but since the 1960s the water body has diminished by 90%—displacing 2.3 million people and leaving 10.7 million people in need of emergency assistance.[7,8] Climate change may also impact the safety of dams. As large amounts of water will come down more suddenly (through one or two storms rather than dozens of smaller storms over the course of a season), the safety of many dams could be in peril as they were not built to accommodate such sudden influxes of water.

As a result of the gradual drying of riverbeds happening on a global scale, an increasingly large amount of power generating dams are becoming less efficient and reliable and are even being shut down. Due to this, new sources of energy need created to replace them and most often these dams are replaced by coal or natural gas power plants, only furthering the effects of global warming. As climate change gets worse, more dams will dry up and could gradually make this technology obsolete.

Because of the limited availability over hydropower, locations of its implementation can be contested, which can sometimes cause international tension. For example, China has heavily dammed some of its largest rivers (such as the Mekong and Yangtze) which has severely destabilized the water supply and raised national security concerns of countries downstream that rely on the same rivers (Laos, Vietnam, Myanmar, Thailand, and Cambodia).

Environmental Impacts. Hydropower has long been hotly debated among environmentalists whether it should be considered good or bad. On one hand, they produce no carbon emissions and are a consistent and economic form of renewable energy. On the other hand, they can have very major impacts on ecology. It is a tradeoff. Both views have very reasonable justification. In some instances, ecology risks can be minimized by location of the dam, but in many cases these impacts cannot be ignored.

Not only do they inhibit the natural travel and migration of fish species, but they can alter the water chemistry in the reservoir and attached water channels. These changes can alter a variety of factors including natural water temperatures, water chemistry, river flow characteristics, and silt loads.[9] Obviously, just like with solar and wind, there is some amount of greenhouse gases created through the construction of the dam. This amount can be substantial but is eventually compensated for by the dam not giving off any emissions.

Except, this might not actually be true. There is some evidence that the reservoirs formed through the use of hydropower are playing an extensive role in the warming of the planet which was previously unforeseen. As these reservoirs sit, bacteria living in the water multiply and feed. The problem here is that reservoirs create perfect habitats for these microbes that would have otherwise not lived. Similar to how humans breathe in oxygen and out CO_2, these bacteria emit the GHG methane (instead of CO_2). This is an especially big issue because methane is much more potent than CO_2. Methane (CH_4) has a warming potential of 28-36 times that of CO_2 over its lifetime in the atmosphere.[10] Because of this, the reservoirs created by dams have had a significant impact on the warming of the atmosphere.

A paper released by Brazil's National Institute for Space Research (INPE) estimated that large sized dams release 104 million metric tons of methane every year.[11] "[Calculations] imply that the world's 52,000 large dams contribute more than 4% of the total warming impact of human activities."[12] Though this means that CH_4 leaks could potentially be substantial, more research is still needed and it is far from settled.

Innovations in Hydropower

Small-Scale Hydropower. In the sight of the problems of increasing unavailability of hydroelectric dams, there has been ongoing research that takes a look at how applicable it may be to create smaller, more modular dams built on smaller flows of water in the hopes that we can expand the influence of hydroelectric power into more areas. Smaller versions of hydropower have massive potential on a global scale, and although many countries around the world do not have rivers large and strong enough to construct a large fleet of effective traditional dams, many do have smaller rivers suitable for small hydropower systems. A 2012 publication that took a look at the future influence that Small Hydropower (SHP) could have in Switzerland, noted that "SHP still has significant potential in Switzerland with the possibility of increasing the production of 2010 by 40-50% by 2050."[13] SHP also has the added benefit of being notably healthier for the environment than full sized dams. Unlike with regular dams, small hydropower often has a proportionally smaller or sometimes no reservoir, which should help the methane emissions situation previously mentioned.

SHP will inevitably be susceptible (potentially even more so than traditional hydropower) to climate changes, so the long-term effectiveness is still uncertain. Many SHP installations will occur in small rivers that are completely dry during some parts of the year, in which case other forms of backup energy will still be required.

Ocean Power. So far, this chapter has spoken specifically on hydropower on rivers, but there is a very promising field that is beginning to emerge that is in relation to producing power from the ocean. This is what is generally referred to as "Blue Energy." Just as green energy generally refers to clean energy on land, blue energy comes from the oceans. One specific type of blue energy refers to the potential energy capture that can occur when a body of freshwater meets the saltwater oceans. When freshwater and saltwater meet, they rush into each other to mix with surprising force. This effect creates huge amounts of pressure in the water where they meet and is fairly consistent and uniform, with the only major adjustment being the tides. By using a certain type of membrane placed where these bodies of water meet, some companies have been able to turn the pressure created here into real energy for us to use. This is generally referred to as Salinity Gradient Power (SGP),[14] utilizing a technology called Reverse Electrodialysis (RED). With the high predictability of ocean patterns and a consistent flow of water out into the ocean, RED has substantial potential. It has the possibility to produce power 24/7, cheaply and abundantly.[15]

Tidal power is another type of emerging blue power, utilizing the same principle as wind turbines, but just underwater. As the current and tides move around under the surface, they create enough motion to turn turbines that are specifically engineered for these conditions.[16] This has shown to be very promising and there are already a couple private companies testing out this technology in the real world. Underwater turbines have also not been shown to have any real impact on the local ecosystem. Because they move much slower than regular turbines and create underwater noise that larger animals seem to know to stay away from, the impact has been shown to be minor and, in some cases, not noticeable, even over a several month period. Similarly, wave power—which uses the motion in the waves on the surface of the ocean to generate power—is making some progress and will be an interesting development to track over the next decade or so.

With vast area around the globe potentially usable for these, we could theoretically power large parts of coastal cities through this energy. Much of it is still experimental, but it has the opportunity to provide a reliable, non-destructive, clean energy source much in the same way dams have, in many more places around the world.

Summary on Hydropower

Hydroelectric power has historically been the largest renewable source used for energy and electricity production. The Three Gorges Dam in China is the largest power production facility on the planet—with a capacity of 22,500 MW.[17] Its capacity is over 2.5 times that of the world's largest nuclear plant, and an even bigger dam of up to a titanic 70 GW has been proposed in the Democratic Republic of the Congo. Now being surpassed by solar and wind because of its availability and environmental concerns, hydroelectric will continue to supply power but is unlikely to grow significantly outside of developing countries. Ocean-based power solutions could change this dramatically, but the technology is still in its infancy.

Chapter 5

Where the Earth is Hot
Geothermal Power, a Limited Near-Future Solution

The Earth is hot below its crust. The immense pressures of mass pressing in on the middle of our planet and the energy from the decay of radioactive materials in the crust creates gargantuan amounts of heat and lava flow below our feet. Some of this heat can eventually find its way back up to the Earth's surface, which is largely how things like volcanoes and geysers are formed. But as much of this that escapes to the surface, there is much more heat trapped just a few miles under the topsoil.

Harnessing this heat for energy production has become a primary focus of efforts to change to renewables. This process is called geothermal energy. Geothermal technology is beginning to make headlines in the renewable energy supply around the world. Because there is heat trapped under the surface everywhere around the world, geothermal plants can (in theory) be built virtually anywhere on the planet if you can drill deep enough. Although technologies are still improving, vast amounts of untapped energy supply are already available to us. As the technology improves, even more untapped energy will become available. In 2019, there was 92 TWh of geothermal power produced; although in order for the industry to keep pace with mitigation climate change, there would need to be 282 TWh produced in 2030—a scenario that would require a 10% annual growth between 2019 and 2030, an extremely unlikely prospect.[1]

Geothermal energy takes the same principle used in traditional power plants, but all the resources are already on the site. Where a coal plant uses the burning of coal and a steam turbine, geothermal uses hydrothermal energy from underground to create energy, often in the same manner. This can be done in a few different ways, but basically steam is either already generated in underground lakes and is pumped up, or water is pumped down pipes underground and brought back up, using the external heat to create steam inside the pipes. Energy can also be created by hot water being pumped up to mix with cold water—when these mix the difference in temperature creates steam.

There are contentions over whether geothermal should really be considered a renewable energy. As we extract steam from underground, the available heat we can utilize will eventually run out or decrease to a level where it is not cost-effective. But as geothermal has become more developed, we find that the steam extracted can be easily pumped back underground (as water) to the source and thus can be run in a sustained manner for very extended periods of time. Furthermore, the sustainability and longevity of geothermal plants has been observed to be immense.

The Larderello Geothermal Plant was the world's first geothermal plant and began operation in Tuscany, Italy in 1913. It has been renovated and expanded several times through its lifetime but amazingly, still operates today. Larderello has survived for more than an entire century and now operates with an energy capacity of 800 MW.[2] This plant has survived almost three times longer than the 40-year life-cycle most standard coal plants are built to last for. Today, the Larderello plant still operates, providing power to around one million homes. Although this site has lasted a very long time, a significant drop in underground pressure has been observed here. The rocks underground are beginning to cool, and the heat is starting to be used up. Eventually, this plant will become inefficient and then obsolete.

There are several other examples of the extreme durability of these geothermal plants such as Wairakei field in New Zealand and the numerous geothermal plant sites in the United States, most notably in California. Although it seems that over time geothermal may not be infinitely sustainable by itself, it is extremely sustainable when compared to other power sources. However, we may over time deplete the currently accessible energy just below our crust and need to dig deeper. Especially with the addition of increased technological capabilities and the recycling of steam into the ground, geothermal may be one of the more sustainable power sources we currently have at our disposal today, comparable to nuclear and hydro. Part of the reason for the long-lasting potential of these plants is simply because it is a relatively simple mechanism. These plants are notorious for the low amount of maintenance that is required for them due to the simple mechanics

One of the great parts about geothermal energy is how effective of a resource it is. Where solar and wind are largely dependent on unpredictable factors (weather, temperature, season), geothermal is very reliable year-round under any weather conditions. In fact, some geothermal plants have a capacity factor of 90%. (To reiterate, capacity factor is a way of measuring how effective a power plant is. It is calculated by taking the actual production divided by the maximum production potential. For example, if a plant ran at 100% power for the entire year, it would have a capacity factor of 100%. Running at 100% for half the year or running at 50% for the entire year would both provide a factor of 50%.) The only other widely used form of power that has a capacity factor this high is nuclear. Even coal plants only run at about 75% availability and some are even lower than this.[3]

For a numerical comparison, according to an EIA report, solar panels in the US will have a projected capacity factor of just 29%, hydroelectric at 55%, wind at 41% and 44% (onshore and offshore), for the year of 2026. This shows the general inconsistencies of comparable renewable energies. In a similar projection, Ultra-Supercritical coal showed an 85% capacity factor (though many other types of coal do not perform this well, with capacity factors as low as 50-60% not being uncommon), while nuclear and geothermal were the highest, showing 90%.[4] (See figure 5) Because of this availability and the fuel being provided naturally, geothermal is one of the cheapest forms of energy. Similarly to other renewables, the costs are high for construction and installation of the plant, but since the costs of maintenance and operation are almost nothing, geothermal is extremely cost effective for consumers. Even further, geothermal plants take up a very small amount of land, relative to other types of sources. Much of the cost for the infrastructure comes from the installation of the underground workings, which can indeed become very expensive. Although solar and wind energies are notorious for having high costs of energy due to their reliance on peakers (see chapter 1), geothermal can provide a more consistent supply of energy for significantly cheaper—with its only real renewable competitor being hydroelectric and potentially offshore wind in the future.

Plant type	Capacity factor (percent)
Dispatchable technologies	
Ultra-supercritical coal	85%
Combined cycle	87%
Combustion turbine	10%
Advanced nuclear	90%
Geothermal	90%
Biomass	83%
Battery storage	10%
Non-dispatchable technologies	
Wind, onshore	41%
Wind, offshore	44%
Solar, standalone[3]	29%
Solar, hybrid[3,4]	28%
Hydroelectric[4]	55%

Figure 5: Capacity Factor By Generation Source
Source: *U.S. Energy Information Administration*. Page 9, Table 1b, "Levelized Costs of New Generation Resources in the Annual Energy Outlook 2021". February 2021 https://www.eia.gov/outlooks/aeo/pdf/electricity_generation.pdf.

Geothermal energy doesn't come without its own problems, however. Because of the way that certain types of geothermal plants extract the steam from underground and then recycle it back, this system can disrupt the underground rock formations. The techniques used in this process are extremely relatable to the fracking techniques used to extract natural gas from underground. Geothermal pipes and fracking both use a technique called hydraulic fracturing (hence the name "fracking") to crack the rocks below the surface, allowing more of the desired substance to flow up (in the case of geothermal, heat). Although fracking is worse than how geothermal plants go about this due to the fact that it uses harsh chemicals instead of water, the geological problems that occur seem very similar.[5]

For example, because of fracking, in 2006 a geothermal energy project in Switzerland was directly tied to the triggering of a small magnitude 3.4 earthquake.[6] The company in charge of the project stopped drilling operations immediately. After operations ceased, the tremors did not continue, and the company paid roughly $6 million in reparations for damage to buildings in the nearby city, which mostly consisted of cracked walls and other minor structural damages. Incidents like these have happened in the past, but most seem to have been mainly because of poor site location, which can unintentionally cause earthquakes that would otherwise have been avoided. In this specific case, the geothermal facility was deliberately built on top of a known fault line. This was done so the heat from underground would be more easily accessible. The company certainly was not trying to cause an earthquake, but the risks were apparent.

The other challenge to geothermal power is location. Geothermal power can be obtained essentially anywhere on the planet if you dig far enough down—as mentioned earlier. However, it is much more efficient in certain geological areas around the globe where there are already large amounts of heat close to the surface. Digging further and further down becomes exponentially more expensive due to the need for proper tools for heat resistance and construction. The shorter the distance needed to drill, the cheaper it is to create an effective power station. Additionally, the closer you are to the surface means a lower likelihood of geological problems occurring down the road. Ideal locations are in regions that are prone to features like geysers, volcanic activity, and where the Earth's crust is generally thinner.[7]

This makes some areas around the world much more efficient and economically feasible than other areas. There is a reason why on the eastern side of the US geothermal is non-existent, but on the western side geothermal is starting to become a noticeable player in the energy sector of many states, namely California and Nevada, but also Oregon, Idaho, Alaska, and Hawaii.[8] One of the most prime regions on the planet for geothermal energy is around the Pacific Ocean's "Ring of Fire." As described in a National Geographic article discussing the significance of this region:

> "The Ring of Fire, also referred to as the Circum-Pacific Belt, is a path along the Pacific Ocean characterized by active volcanoes and frequent earthquakes. Its length is approximately 40,000 kilometers (24,900 miles). It traces boundaries between several tectonic plates—including the Pacific, Juan de Fuca, Cocos, Indian-Australian, Nazca, North American, and Philippine Plates. Seventy-five percent of Earth's volcanoes—more than 450 volcanoes—are located along

> the Ring of Fire. Ninety percent of Earth's earthquakes occur along its path, including the planet's most violent and dramatic seismic events." [9]

This region of hydrothermal activity on the planet is likely going to be the largest area of geothermal expansion that we can expect to see in the near future. Other countries around the Pacific Ring of Fire such as Japan and Indonesia will be able to meet a large amount of their energy consumption with geothermal. Both of these countries have large amounts of hydroelectric capacity, so with the addition of geothermal it seems very probable that these countries could provide for large amounts of their needs with renewable energies.

There are of course regions outside of the Ring of Fire that can be good for geothermal. For example, in 2018 geothermal produced an impressive 44% of Kenya's total electricity, and 30.8% of Iceland's.[10,11] (In the case of Iceland, another 69% came from hydro, with the remaining 0.2% coming primarily from wind power.)

Summary Of Geothermal

Outside of these regions of high hydrothermal activity is where the problem lays. Although we can access geothermal power if we drill deep enough, it does not show competitive prices compared to other fuels when it isn't close to the surface. The deeper you have to drill the more expensive the project becomes. Because of this, large areas around the planet are just not economical to produce this type of power yet. For this reason alone, geothermal is an improbable solution to our energy crisis by itself. It can, however, certainly provide a helping hand. It would be unfair to say that the technology could advance enough to allow for more competitive pricing in every location across the globe within the next couple decades because it simply won't be. The deepest humanity has ever drilled below the surface is the Kola Superdeep Borehole SG-3, which dug 12,261 m (about 7.6 miles) under the surface[12] This might sound deep, and it was, but it was also incredibly expensive, not to mention that it took 24 years. Even further, the hole was a mere 9 inches in diameter.[13] Knowing this, we are very far away from being able to provide economically feasible geothermal energy to many parts of the world where it is not currently easily accessible. One of the final benefits to geothermal is that since the technology can be used in ways that are similar to that of natural gas fracking, many workers from that industry can be retrained to help with geothermal.

In 2020, geothermal provided only 0.4% of total US utility-scale electricity generation (which was still over three times the amount produced in Kenya).[14] While the US is the world's largest producer of geothermal energy, there are still multitudes of untapped potential and a long way to go.

Section II:

Our Friend the Atom

Chapter 6

The Solution We Hate
Introduction to Nuclear Power

It was early morning in the deserts of New Mexico. A team of scientists and military personnel huddled around encampments dotted across the open sands, watching the sun steadily rise into the morning of July 16, 1945. In an instant, the sky was filled with fire and destruction seen at a scale thousands of times larger than the likes of which has ever been seen before in human history. The atomic bomb had been created. Physicist Robert J. Oppenheimer, the so-called "father of the atomic bomb," stands witness to the new age of mankind that he has brought onto this planet, the nuclear age. Oppenheimer is remembered famously at this moment for reciting a Hindu scripture in saying, "Now I am become death, the destroyer of worlds." The truth is, the knowledge gained from Oppenheimer's work in creating the atomic bomb would come to help mankind in profound ways, rather than destroy it.

The world desperately needs a steppingstone between the destructive fossil fuel burning of the present, and the utopian 100% clean energy of the future. Not only will the continued use of fossil fuels continue to damage the planet, but eventually the resources will be depleted and run dry. If we are still dependent on these resources when that time comes, the consequences could be disastrous. In the case of wind and solar, the energy provided is not consistent and cannot yet be supplied effectively to many areas of the globe. Hydroelectric and geothermal are both renewable and consistent but have limited availability. So the challenge is creating a consistent form of energy that does not create greenhouse gases or pollution and can be implemented anywhere. This is where nuclear power has a part to play.

Nuclear power is one of the most controversial topics concerning climate change and energy generation. An annual poll conducted by Gallup News revealed that as of March 2019, 49% of Americans favored nuclear power and 49% opposed it.[1] Over the years, nuclear power went from one of the most accepted and positively viewed forms of power to one of the least favorable. One study even found that US opinions on nuclear power in 2016, on average, were close to opinions on coal mining, fracking, and offshore drilling.[2] Granted, it's understandable why there is so much stigma associated with nuclear energy.

Not only did it come from insidious nuclear weapons, but terrible incidents like at Chernobyl, Three Mile Island, and Fukushima—which will leave large amounts of land simply uninhabitable, not to mention the thousands upon thousands of deaths that were caused—are simply unavoidable. There are massive amounts of toxic, radioactive waste that will essentially never go away. Nuclear power gives countries an easy way to produce nuclear weapons, raising major global security concerns. Radioactive waste obtained by terrorist organizations could be used to make a so-called "dirty bomb." Not to mention that nuclear power is simply not economical anymore, or that uranium mining is incredibly damaging to the environment. And of course, the nuclear-industrial complex is "sketchy"—to say the least—and is not concerned with these issues, but rather is much more focused on maximizing profits. There is simply no justifiable reason today to rely on nuclear power.

At least, these are the arguments that opponents of nuclear power use. The truth is, almost all of these arguments are simply false, as we will see throughout this section.

That's the concerning thing, because this research makes it clear that to at least half the population, nuclear power has very major drawbacks. This is very contradictory as to the actual science and facts behind nuclear power. The large majority of scientists and researchers who know much about and deal with nuclear power see nuclear as either one of the (if not *the*) best options for producing energy, or that it must at least have a substantial part in the energy mix of the future to provide an emissions free baseload power supply to support renewables. Despite this, it is not treated proportionally. Especially when compared to fossil fuels, nuclear has clear advantages. Just the facts that 1kg of natural uranium fuel for a nuclear reactor produces 12,500 times more energy than 1kg of coal[3] and a single uranium fuel pellet (1/2" height & diameter cylinder) contains the same energy as 149 gallons of oil (and a typical reactor holds up to 18 million fuel pellets at any given time)[4] should be enough to persuade more people for the use of this technology, but that isn't what we see happening today.

Additionally, nuclear has the potential to be an immense economic benefit for the future of not just the United States, but around the globe. The US nuclear industry directly employs almost 100,000 people in high quality wages, and nearly 475,000 indirectly.[5] On average, nuclear worker salaries are 20% higher than those of other electric generation sources.[6] To be fair, a shift to renewable power in the US would create more jobs than increased incentive on nuclear power, but they also come with the drawbacks mentioned in previous chapters. Nuclear power is also the reason why some countries, such as France, have such low electricity cost rates.

Because of the real benefits met with large controversy, the next several chapters are all dedicated towards mainly debunking myths about nuclear energy and informing about upcoming nuclear research that is very commonly overlooked but may give way to highly advanced sources of energy. Nuclear power certainly has its drawbacks, but it is crucial to understand the line between reality and hysteria, especially as it may be one of our most viable options for fighting climate change.

Disinformation of Nuclear Power

The running joke is that the nuclear power industry "has the best scientists and the worst public relations team." This unfortunately has been disastrously accurate, and has caused the industry much pain. In the face of confrontation, the nuclear industry has historically spent much of its resources on things such as publicly accessible information and fact sheets instead of the typical lobbying of politicians (though this does occur as well) while disinformation campaigns led by other industries—namely various fossil fuel corporations—have been massively successful at turning the public against nuclear. Instead of getting on the nightly news, "Big Nuclear" would instead make a fact sheet and put it on its website. This strategy made the industry extremely vulnerable to attack, with no effective method for rebuttal. Fossil fuel corporations invested heavily in climate change disinformation—as has been thoroughly reported by a legion of institutions, but the Union of Concerned Scientists has done an exceptionally detailed job[7,8,9]—and so it should not be surprising to hear that these same companies also invested heavily in disinformation against nuclear power (which, until recently, was the only serious competitor against fossil fuels).

The reasoning behind this is actually quite simple; if fossil fuel interests can reduce the use of nuclear power, they have less competition. These interests don't believe (or at least, have not historically believed) that wind and solar will make a large impact on energy dispersal, so therefore nuclear power is the only serious competitor. (Though obviously these companies would still prefer for renewables to not be implemented.) Even if renewables are implemented at a large scale, they will still allow for fossil fuels to exist through the use of peakers. A common theme that will be explored throughout this section will be that when nuclear plants are shut down, they are often replaced by coal or natural gas plants, *not* by solar or wind. "But how extensive could fossil fuel, anti-nuclear disinformation really be?" you may ask. Shockingly extensive.

What qualities does Big Oil possess? Environmentally destructive? Check. Unsympathetic? Check. Greedy? Check. However, Big Oil is nothing if not intelligent and strategic. This is likely why much of the money spent from Big Oil on anti-nuclear activities went to three places—lobbying, typical disinformation advertisements, and donations to environmental groups. You read that right. Fossil fuel corporations have historically been some of the largest donors to environmental groups. Predictably, so have renewable energy interests. Therefore, it is a one-two-punch against nuclear. There are too many examples to go over for the scope of this book, but we can look at a few just to have a taste.

This scenario has caused what Environmental Progress (a strong pro-nuclear organization) calls "The War on Nuclear." [10] To take a few examples from them, the well-respected Sierra Club has "taken $136 million from [natural gas]/renewables interests that stand to profit from the closure of its nuclear plants" (the majority of this does come from renewable backing). The Natural Resources Defense Council (NRDC) has a "minimum of $70 million directly invested in oil and gas renewable energy interests that stand to profit from the closure of nuclear plants." The Environmental Defense Fund (EDF) has received a "minimum of $60 million from oil, gas, & renewables investors who would directly benefit from EDF's anti-nuclear advocacy."

As Forbes reports, the American Petroleum Institute has "flooded the airwaves in Ohio and Pennsylvania with anti-nuke commercials by pushing fear – fear of higher prices and fear of radiation. Just the opposite of what is true. …This is ironic since the natural gas industry emits more radioactivity than all of nuclear in America combined. And more people <u>die every year from natural gas</u> than any other electricity source except coal." [11] And, "In 1970, a leader of the petroleum industry and the head of the Atlantic Richfield Co. named Robert O. Anderson contributed $200,000 to fund Friends of the Earth, an organization that is strident in its opposition to nuclear energy." [12]

> "For several years following the issuance of the [first Biological Effects of Atomic Radiation (BEAR)] report, committee members and various Rockefeller Foundation-supported researchers worked hard to spread the idea that there are no safe doses of radiation. Their effort helped to fertilize the seeds of doubt about nuclear energy that the committee carefully planted. …The growing apprehensions contributed to a number of expensive regulatory requirements and set the stage for a focused movement to oppose all efforts to develop nuclear energy." [13]

As we will see in chapter 8, this idea that there is no safe dose of radiation is simply not true by any perspective. But for now, chew on the fact that eating a single banana exposes you to more radiation than living next to a nuclear plant does in an entire year.[14] *

In the hit seller *This Changes Everything*, Naomi Klein (who is strident in her opposition to nuclear power throughout the book) highlights this partnership further:

> "For instance, Conservation International, The Nature Conservancy, and the Conservation Fund have all received money from Shell and BP, while American Electric Power, a traditional dirty-coal utility, has donated to the Conservation Fund and The Nature Conservancy.

---

* Bananas contain potassium-40, a radioactive isotope of the element. The radiation emitted can be significant enough to set off some sensitive radiation alarms, though obviously it is not enough to hurt you. A common unit of measurement for radiation dosage is the BED- Banana Equivalent Dose.

> WWF (originally the World Wildlife Fund) has had a long relationship with Shell, and the World Resources Institute has what it describes as 'a long-term, close strategic relationship with the Shell Foundation.' Conservation International has partnerships with Walmart, Monsanto…BHP Billiton (a major extractor of coal), as well as Shell, Chevron, ExxonMobil, Toyota, McDonald's, and BP. …And that is the barest of samplings."

Big Oil has both directly and indirectly targeted the nuclear power industry with its disinformation campaigns for decades, and this has ironically been the root cause of many of the anti-nuclear power movements that align themselves so much with the renewable power movement. In one leaked letter from inside the fossil fuel industry, nuclear waste storage was on the list of what should be the target of a widespread disinformation campaign, second only to climate change science itself. Big Oil not only funded wide-scale disinformation about nuclear power, but it also—quite successfully—played both sides of the political isle and took advantage of every social, economic, and political aspect that it could. It is no coincidence that many of the arguments still used against nuclear power to this day by "environmental organizations" (i.e., large waste amounts, unsafe, poisons local populations, waste cannot be effectively stored, leads to more nuclear weapons, inherently unsafe) originally came from fossil fuel interest-funded propaganda.

Throughout this section it will be important to keep an open mind. The topic of nuclear power is extremely polarizing, and the only way to fairly assess the realities behind it are to forget any pre-conceived conceptions you may have (both for and against it) and hear out the perspective laid out in the coming chapters. Despite nuclear having a shaky history, it will be an essential resource in order to rapidly and effectively decarbonize the energy sector of the world. Because of its massive role to play in the face of massive opposition, it feels justified to discuss nuclear power extensively in this book in order to clear misconceptions about the technology.

Chapter 7

The Great Misconception
Current Nuclear Power, a Limited Modern Solution

Though there are two separate types of nuclear reactions, for most of this section we will focus on nuclear fission (though chapter 9 talks about the other type—nuclear fusion). Nuclear fission is the form of nuclear power that we use today at all nuclear power plants, including at the Chernobyl, Three Mile Island, and Fukushima disasters. Through the process of nuclear fission is how the first atomic bombs and reactors alike were created. It is raw power. Understanding the process of nuclear fission helps to comprehend why we can make both massive bombs and stable power from it. Put into simple terms, nuclear fission works as such:

1. An atom (typically of Uranium or Plutonium) is first shot with a neutron, causing it to split into a smaller element and release more neutrons (often releasing multiple neutrons from the source atom). This process releases energy in the form of heat and creates waste products. (See figure 7.1)
2. Neutrons then fly from the parent atom and hit other atoms, which then causes more neutrons to be released and so on, creating a chain reaction.
3. This process will continue forward until the neutrons are stopped by an outside force or until the reacting material is used up.

A simplified explanation of the difference between weapons and energy is derived from two key differences. First, our ability to "regulate" a nuclear reaction in a reactor (by slowing down or absorbing neutrons) is beneficial for producing energy, whereas this reaction is better left to its own without regulation in a weapon so that it can release energy rapidly. Second, a nuclear bomb uses a much higher concentration of fissile (meaning able to undergo a fission reaction) material. In a nuclear reactor, only 3-5% of the material in the reactor is fissile. In a bomb,

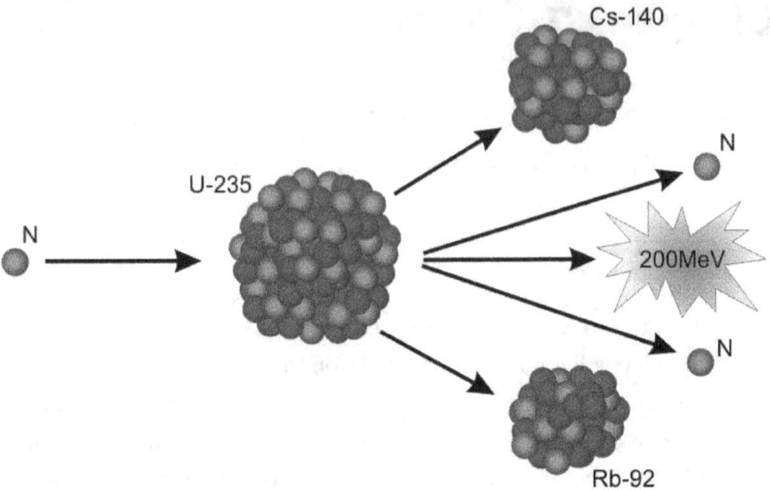

Figure 7.1: Corefission of Uranium-235
N stands for neutron. Source: Wikimedia Commons. The waste products depicted here are Caesium-140 and Rubidium-92, though various other wastes are created during the fission process, especially when considering waste generated through radioactive decay.

this figure is typically 90%. If a nuclear weapon were attempted using the material composition of a nuclear reactor, simply nothing would happen. In a way, comparing nuclear power to nuclear weapons is like comparing trees to coal (since coal is just dead plant matter that was concentrated and condensed over millennia).

Nuclear Power Basics

To start, nuclear power plants work the same as most others: creating steam with heat and using the steam to turn a generator, creating an electrical current. To regulate the heat, control rods are used. These rods act like a sponge, absorbing and holding onto roaming neutrons in a reaction. In the case of a runaway chain reaction, more rods can be added and/or inserted deeper to slow or completely halt the fission process.[1] Since these rods still leave the reactor hot because of the heat that had already been generated, cooling water is used to cool down the reactor (and also helps to further regulate the chain reaction by slowing down the neutrons which increases their chances of initiating fission with the uranium, increasing operational efficiency).

The most commonly used type of nuclear reactor is the Water Cooled Reactor (WCR, though designs vary widely within this category), which uses water to prevent heat from rising too high in the reactor. This not only allows the temperature to stay at safe levels but also allows for the reaction to keep going. More than 95% of all operating civilian power nuclear reactors are of this type, and the most common types of this reactor are the Light, Pressure and Boiling Water Reactors (LWR, PWR & BWR).[2] The WCR is also the generic type of reactor that was involved at Three Mile Island, Chernobyl, and Fukushima.

Nuclear reactors are also used extensively on military vehicles. US aircraft carriers and submarines are powered by nuclear power. For example, the Gerald-Ford aircraft carrier costs $11.4 billion (although price overruns have taken this to $13.3 billion for the first model) and is powered with two A1B nuclear reactors. Using fossil fuels for these types of vessels does not makes sense because of the massive amounts of fuel that would need carried onboard, and the frequency of which these ships would require refueling, as well as other factors. Today, nuclear power provides around 10% of the worlds electricity.

Modern Safety

If the control rods are not inserted to the reaction to stop it, the reaction will continue and eventually can grow so hot that it will melt through the protective walls containing the radiation, or cause an explosion, leaking it into the open.[3] This is *incredibly* rare and has only ever occurred primarily due to human error and shortcutting/short sightedness in the design and construction of nuclear facilities—which is exactly what happened at the Chernobyl and Fukushima meltdowns, which are discussed in depth in chapter 8.

Keeping these in mind and factoring in other safety mechanics, modern nuclear energy is actually an incredibly safe form of energy production. Even if you include disasters, nuclear energy is by far one of the safest forms of energy. In reality, the Chernobyl incident has been the only nuclear incident "in the history of commercial nuclear power that radiation-related fatalities occurred." [4] This shows more than just that nuclear plants are safe; it shows that non-nuclear forms of energy generation can be very unsafe—specifically fossil fuels.

You don't often think about fossil fuel waste management because, well, most of the waste just goes into the air. This is commonly thought of as CO_2 and other greenhouse gases, but it is a lot more than that. It is exactly because of this reason that nuclear and renewables are so much better for our health than fossil fuels. Every human on this planet actively breathes in fossil fuel waste on a day-to-day basis, with drastic impacts on our health. It is common knowledge that fossil fuels are terrible for the environment, but what these industries don't want you to know is really how bad it is for you as an individual. Fossil fuel emissions can (and do) literally kill people, both directly and indirectly. As emissions continue to impact the atmosphere, global warming, climate change, and risks to human health will continue to grow as well. Not to mention the thousands of miners that are killed directly every year (especially from coal mining). (See chapters 18 & 19.)

There is extensive evidence of various products of emissions being linked to heart and lung cancers and diseases. The Centers for Disease Control and Prevention (CDC) says that heart disease alone causes approximately 647,000 deaths (1 in every 4 deaths) in the US every year, and cost the US $219 billion in both 2014 and 2015.[5] It turns out that heart disease is also one of the most highly established health risks associated with the burning of fossil fuels.[6] I am not saying that everyone who dies from heart disease is a victim of the burning of fossil fuels, just that there is a strong correlation between fossil fuel emissions and health risks that is not present with nuclear and other forms of energy and that many who die prematurely of a plethora of diseases do so because of the impacts fossil fuel emissions have had on their health. This is especially prevalent in those living in urbanized areas.

Obviously, nuclear power is not completely free from GHG emissions—just like renewables. Also like renewables, they pay off their CO_2 debts over their lifetime. The emissions coming out of nuclear power cooling towers are not CO_2 or smoke, just water vapor. According to a 2013 study looking at the difference of harmful emissions between nuclear and fossil fuels, nuclear power was shown to prevent "...about 15 times more emissions than it caused." The study continued to state that:

"Using historical electricity production data and mortality and emission factors from the peer-reviewed scientific literature, we found that despite the three major nuclear accidents the world has experienced, nuclear power prevented an average of over 1.8 million net deaths worldwide between 1971-2009. This amounts to at least hundreds and more likely thousands of times more deaths than it caused. An average of 76,000 deaths per year were avoided annually between 2000-2009 [(see figure 7.2)], with a range of 19,000-300,000 per year." [7]

It is because of this fact that there are no post construction emissions that renewables and nuclear power are so much safer than their fossil fuel competition (in addition to safer mining conditions, despite the radiation). The most harmful part about the nuclear process is the nuclear waste created as a byproduct, but arguments going against this nuclear waste are typically driven by fear and opinions instead of facts and rational thinking.

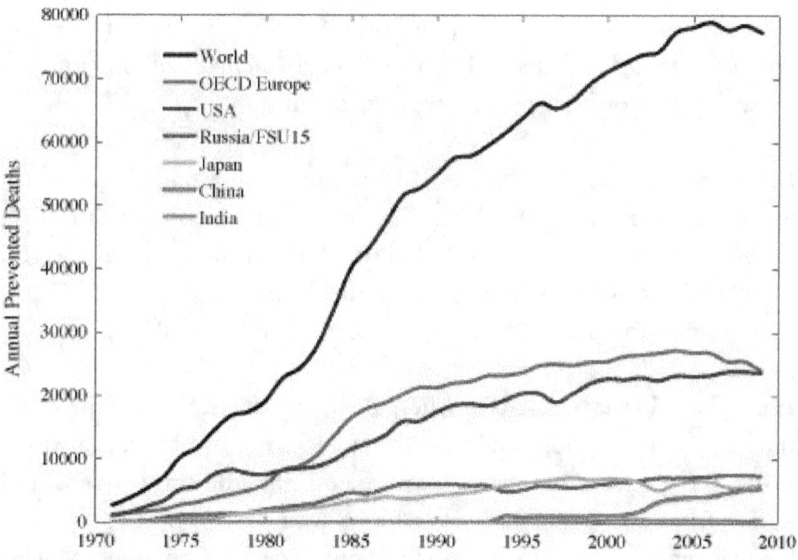

Figure 7.2: Mean number of deaths prevented annually by nuclear power 1971-2009

Source: https://pubs.acs.org/doi/10.1021/es3051197, Pushker A. Kharecha and James E. Hansen, *Environmental Science & Technology* 2013 *47* (9), 4889-4895. Taken with permission by the American Chemical Society.

Nuclear Waste and Radiation

Yes, nuclear reactors produce long-lived radioactive waste, but it is not nearly as much of a problem as it is made out to be. There are four basic categories of nuclear waste: Very Low-Level Waste (VLLW), Low-Level Waste (LLW), Intermediate-Level Waste (ILW) and High-Level Waste (HLW). (LLW and VLLW are commonly both referred to together as LLW.)

Very Low-level Waste. VLLW is not considered harmful to people or the environment. It is comprised primarily from demolished materials such as concrete and piping. Other industries that produce this type of waste are the steel, chemical, and food processing industries.[8] A significant amount of nuclear waste is of this type.

Low-Level Waste. LLW is short-lived and not strongly radioactive. It doesn't require shielding and can be easily handled. LLW is comprised of paper, rags, tools, clothing, and other materials. It is produced by not just the nuclear industry but also the medical field and other industries.[9] The majority of nuclear waste is of this type.

Intermediate-Level Waste. ILW requires some shielding, but still is not a large hassle for handling. It is not extremely hot or radioactive, but some ILW may be long-lived. This comes from chemical sludges, fuel claddings, and contaminated materials from reactor decommissioning.[10] Some long-lived ILW may be stored with the HLW.

High-Level Waste. HLW is where the concern lays. This is the waste that is commonly referred to when discussing issues with nuclear waste. HLW is radioactive enough to require both shielding and cooling and comes from spent fuel and other highly radioactive parts. Once the waste is removed from the reactor, it is put into cooling pools for at least five years and then can be put into dry storage casks (that are kept on-site) once sufficiently cooled. Afterwards, it can eventually be put into heavy steel drums for long term storage. Since the fuel is solid—and not liquid as many think—leaks from this are not nearly as much of an issue as many assume.[11] A very small amount of nuclear waste is of this type, but stays radioactive for up to several hundred thousand years.

Thankfully, the *vast majority* of nuclear waste is LLW. "Most of the radioactivity present in radioactive waste (up to 95% of the total) is present in HLW (including spent fuel, when declared as waste). In terms of volume, it is the reverse, and more than 95% of the total volume of waste comprises LLW or VLLW." [12] Less than 1% of nuclear waste by volume is HLW. (See figure 7.3)

Additionally, large amounts of VLLW and LLW have been disposed of, while no HLW is disposed of due to its long-lived radioactivity. To put the global volume of HLW in storage in perspective, it "is a volume roughly equivalent to a three [meter] tall building covering an area the size of a soccer pitch." [13] In short, this means that the amount of radioactive waste created that requires long-term storage is incredibly small. While there are huge amounts of liquid HLW in storage (~2,786,000 m3), this has mostly been created from defense activities and weapons testing, not from nuclear power plants for energy production. The HLW produced by nuclear power plants is almost entirely

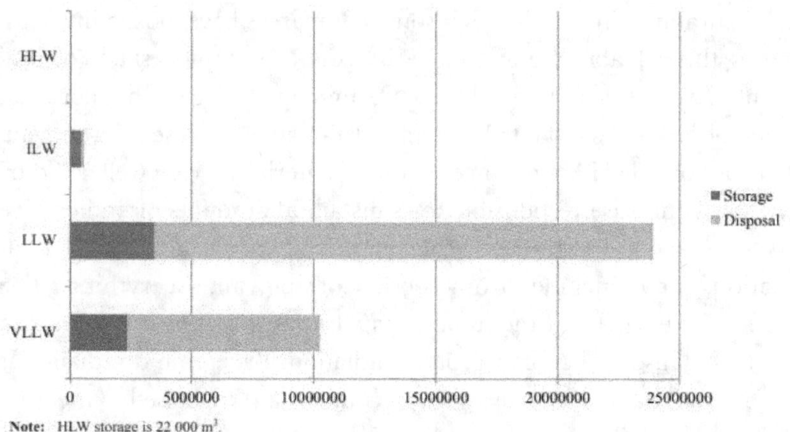

Figure 7.3: Summary of global solid radioactive waste inventories (m3)
HLW is 22,000 m³ (all in storage), ILW is 460,000 m³, LLW is 3,479,000 m³, and VLLW is 2,356,000 m³. Source: INTERNATIONAL ATOMIC ENERGY AGENCY, Status and Trends in Spent Fuel and Radioactive Waste Management, Nuclear Energy Series No. NW-T-1.14, IAEA, Vienna (2018). See citation "12" this chapter for more details. Taken with permission by the IAEA.

solid material. So, while *The Simpsons* loved to depict huge barrels of radioactive leaking green goo, this is not the reality, but it is understandable how this narrative was created. A site in Hanford, Washington is commonly used to argue against the storage of nuclear waste, but the reality is that almost all of this waste is from military projects.

To compare this to the amount of waste produced by the coal industry, the Nuclear Energy Institution (NEI) again reaffirms that, in the US, "coal plants generate that same amount of waste every hour." [14] These figures are mind-blowing, but it is important to take this point with a grain of salt. It would be misleading to not discuss that this "soccer field volume" is not the same size as it takes to contain all of the waste, as there needs to be facilities to deal with it. It also must be pointed out that coal has always been and is still currently used far more commonly than nuclear energy. In 2018, coal produced 38% of the world's electricity and nuclear provided just 10%. [15]

Despite this, you cannot look over the fact that the waste from other power sources isn't radioactive—like with nuclear. Except, this isn't exactly true either. When coal burns, it releases all sorts of things besides CO_2, and some of these are slightly radioactive. Fossil fuels release so much uranium into the air that some companies have looked into capturing this valuable energy source right out of the smokestacks of coal plants. Even aside from fossil fuels, the mining of rare earth metals and minerals releases substantial amounts of radiation into the environment. Because of solar PV's reliance on these materials, it very well could be true that they are responsible for substantial amounts of radioactive waste, as are all electronics. In fact, you are always being exposed to radiation. That includes from food, clothing, air. Everything. The difference here is that the radioactivity on most matter is too small to hurt us. This is called background radiation. Background radiation is the reason that if you take a Geiger counter (a device used to measure radiation), it will always give you a slight reading.

Not only that, but you are exposed to more radiation depending on what elevation you're at. Taking a single plane flight exposes you to hundreds of times the radiation received from living near a nuclear plant for a year. It is significant enough that one study found that flight attendants "had a higher prevalence of every cancer we examined, especially breast cancer, melanoma, and non-melanoma skin cancer among females." [16]

Flight crews are even classified as "radiation workers".[17] Someone living in the mountains of Colorado is exposed to a higher background radiation dose than someone living at sea level, on a daily basis.

Back to coal though, when a coal plant burns tons of coal every year (literally), this radioactivity can accumulate in a very profound way. It is so profound that coal plants operate well outside of the allowable range for radiation that nuclear plants are held to. As put by an article in Scientific American, "In fact, the fly ash emitted by a power plant—a byproduct from burning coal for electricity—carries into the surrounding environment 100 times more radiation than a nuclear power plant producing the same amount of energy."[18] The amount of radiation leaked from nuclear plants (foregoing a disaster) is so small that you are exposed to more radiation from the granite countertops in your house than if you live near a nuclear plant.[19] And as discussed in chapter 1, PV solar waste also has its own issues with heavy metal and electronic waste, which can be extremely toxic in the end-of-life stages. Natural gas creates extensive toxic (and slightly radioactive) waste at the site of extraction. Electric vehicles and battery storage will create mountains of electronic waste. The point is, everything humans do creates hazardous waste. It is only a problem if not treated with care.

A misconception that people who are pro-nuclear commonly portray is the argument that "Nuclear waste is around for a long time, sure. But heavy metal waste from solar panels and other electronics or chemical waste from industrial processes never goes away because it is not radioactive. If anything, nuclear waste is preferable to the infinite toxicity of these elements." The problem with this argument, though scientifically correct, is that most nuclear decay chains leave the final element as lead, which is obviously very toxic in of itself. In the end, nuclear waste becomes a toxic element just like most other wastes. The difference is that the nuclear waste is handled more responsibly.

If not handled properly or left out exposing the environment, there could most certainly be large amounts of radiation contaminating the air, soil, and water supplies from nuclear waste. The containment of the LLW is extremely easy, with governmental policies often simply storing it in shipping containers and then into landfills if/when applicable. ILW does require some storage, but afterwards it can be stored in specified containers. While HLW only comprises of a small amount of the waste, it needs the most protective storage. As discussed previously, this involves leaving the waste in pools of water for a few years until it cools down enough to store the waste on-site in dry-casks. These containments work so well that you can go right up to the wall of the casks and touch them without any protective gear and be perfectly okay. It is also important to understand that (generally) the longer-lived the waste is, the less radioactive it is. The most radioactive substances decay very quickly. So while it may be around for a long time, most HLW is not extremely radioactive (per atom) when compared to some other radionuclides. This is why the reasoning about nuclear waste being around for so long is a little misleading. If the only concern was about how long it will be around, we would dig up all of natural uranium in the Earth and run it through a reactor, since natural uranium has a half-life of billions of years. The concern isn't actually how long it will be around for. It is a psychological problem because it is seen as a foreign substance, unnatural, and dangerous. This perspective is far from reality.

The larger issue is a matter of figuring out where to store the HLW in the distant future, when larger amounts of material are built up. Places like Yucca Mountain in the US and Finland's Onkalo site are being considered for long time storage, though there are a host of problems with storing nuclear waste over thousands of years. These include not just security and environmental safety but also changing geology and debates on how to designate the site as a hazard to civilizations far in the future who might not even speak the same language. Consider the possibility of a cataclysm that sets humans back thousands of years—such as a nuclear war or devastating cosmic impact. In this case, it is extremely likely that the surviving humans who repopulate the planet (likely after hundreds or thousands of years) will speak a completely new language and have no understanding of nuclear waste or radiation. How would you signal to these future humans that this area is a serious hazard? How would they interpret different signs?

Again, this is a problem that is important, but also receives a massive overreaction. While this should be a priority, the current inventory of HLW is, again, exceedingly small and is not an imminent threat. Additionally, future nuclear fission technologies (see chapter 10) will allow us to burn existing nuclear waste again and reduce the amount of waste in our inventories to hundredths the current volume. Energy sources such as nuclear fusion will create (for all intents and purposes) zero nuclear waste. Because of this, storing waste long-term is a minuscule and distant problem in reality. As put by the World Nuclear Association, "Nuclear power is the only large-scale energy-producing technology that takes full responsibility for all its waste and fully costs this into the product," and the industry will only continue to improve upon this.[20] Nuclear waste is seen as an issue that is inherent to nuclear power, but in some ways the opposite is true. HLW is generated in part because the nuclear power used today was not optimized for power generation but was optimized to specifically produce plutonium that could then be used again for nuclear weapons (more on this in chapter 10) with power generation as an afterthought. A nuclear power source that is designed specifically for power generation from the ground up will produce little to no waste, and the waste that is created will be more difficult to create weapons from.

There have also been a couple of high-profile nuclear waste disasters that are still ongoing worth discussing such as the increasing concerns at the Marshall Islands and Church Rock, New Mexico—both of which are blood on the hands of the US. (The nuclear disasters at Three-Mile Island, Chernobyl, and Fukushima are each discussed in detail in chapter 8.)

Marshall Islands. The Marshall Islands nuclear waste situation is frequently pointed to as a reason for the dangers of nuclear power. The big misconception here is that the waste is not from nuclear power, but from nuclear weapons tests conducted by the United States.

During the 1940s and '50s, the US tested nearly 70 nuclear weapons in the Marshall Islands.[21] This in turn released large amounts of radioactive fallout across the islands and created significant amounts of radioactive waste. One specific explosion on Runit Island—a small, sandy island barely above the sea—created a crater that was determined to be used as a storage site for radioactive waste. Because of this, massive amounts of waste from tests in the region (as well as waste from stateside nuclear and biological weapons tests) were dumped into the pit, as the *Los Angeles Times* reported.[22] After the material was dumped, a concrete dome was constructed overtop to seal the radiation and waste inside.

However, in an act of spectacular incompetence, the US military did not line the bottom of the crater with any concrete or other type of containment. Now, the waste is possibly leaking out through the porous coral rock of the island and into the ocean. Adding fuel to the fire, the top of the containment is also now cracking.[23] As sea levels rise, there is also potential for the sea to come into direct contact with the cracking, aging cover, as the island barley sits above sea level. This could obviously raise enormous concerns about radioactive fallout. Today, the nuclear dumpsite at Runit Island is a major nuclear disaster waiting to happen, and the future of the Marshall Islands is far from safe.

Church Rock, New Mexico. Just a few months after the Three Mile Island nuclear accident, another nuclear incident occurred inside of Navajo Nation. The plethora of uranium mines around the region created large amounts of contaminated waste (mining gear, uranium, contaminated water and sludge) that was then stored in disposal ponds. On July 16, 1979, one of these ponds in Church Rock, New Mexico cracked open and leaked "approximately 1100 tons of radioactive mill waste and 95 million gallons of mine process effluent" [24] into the nearby Puerco River (often referred to as the "Perky" by the indigenous community), which the indigenous community relied on heavily. Further mismanagement of this site continued to contaminate the groundwater and surrounding environment.[25] The spill was the largest release of radioactive waste by volume in US history and was one of the largest releases of radiation in world history.[26]

In true western culture fashion, the indigenous community was not given proper warning about the risks of the leak on public health. This is true not just for the leak at Church Rock, but also for the health implications of those working in the uranium mines (which employed many indigenous members). Cancer rates in these communities have skyrocketed in the past several decades. A study published in the Journal of Occupational and Environmental Medicine found that between 1969 and 1993, two-thirds of all new lung cancer (which can be caused from breathing in radon and uranium in mines that are poorly ventilated) in Navajo men were attributable to underground uranium mining.[27] Furthermore, the same study found that the risk of lung cancer among male Navajo uranium miners was a staggering 28 times higher than

Navajo men who did not mine. Though the true health impacts are not well understood and there is limited information, this has obviously had a serious impact on the affected communities.

While cleanup is underway, the area is to this day still contaminated, and the EPA recommends members of the community relocate outside of Navajo territory until the cleanup is complete.

Uranium Mining. The mining of nuclear power's primary fuel, uranium, is commonly pointed out as a major drawback to nuclear power. As mentioned, the indigenous community of Navajo Nation has suffered from poor uranium mine conditions, and there are a plethora of other impacts. Open-pit mines in particular release large amounts of radon, which can be hazardous to local populations and ecology. While uranium mines in developed countries such as the US, Canada, and Australia are of minimal concern to worker health and environmental damages because they are highly regulated (despite the Church Rock incident), they are much more of an issue in less developed countries such as Kazakhstan (who produces over 40% of the worlds uranium[28]), Namibia, Uzbekistan, Niger, and Ukraine, where regulations—both for worker and environmental safety—are more relaxed. Additionally, the radiation from the uranium itself is low (similar to granite), but the most dangerous radioactive release comes from radon.[29] Natural uranium gives off so little radiation that you can buy it on Amazon.com, as I have.

Uranium mining is consistently pointed out as one of the most environmentally damaging materials to mine for, but it is in actuality very comparable to the mining of other heavy metals. Additionally, it is important to keep everything in context. Are a handful of uranium mines really worse than thousands of slightly less radioactive coal mines? Because of the sheer minimal amount of material needed, uranium mining has significantly less of a net-ecological impact than something such as coal or natural-gas fracking, the heavy metals that need mined for solar power and electronics, and the massive ecological displacements of hydropower. If you are going to argue against nuclear energy on the basis of uranium mining, it is important to reevaluate most green technologies, because they have mining damages that are substantially worse than uranium.

There just isn't an easy answer for this topic. The reality is that everything humans mine has significant environmental impacts. In the case of nuclear power specifically, nuclear fusion would make this issue essentially non-existent (see chapter 9) and nuclear fission that uses a thorium fuel-cycle could greatly reduce the impacts as well (see chapter 10).

Other Considerations for Modern Nuclear Energy

Despite the increasing construction and engineering costs, nuclear power plants can still be economically viable. France is the world's leader via percentage of total power produced coming from nuclear energy, (producing 71.7% of total electric generation in 2018, though this has declined in years since)[30] and the benefits are clear. Because of this, French citizens on average pay some of the lowest electric bills on the planet. In 2019, "the average cost of electricity in France [was] 26.5% cheaper than the EU average, compared to Spain or Germany where prices are respectively 46% and 79% higher than France's." [31] On the contrary, as mentioned earlier in the book, Germany and California have both invested extensively into renewables (particularly solar and wind power), and because of this they pay some of the highest electric costs in the world. It has been suggested by some sources that if Germany and California had invested their combined $680 *billion* into nuclear power instead of renewables, they could both be producing more than 100% of their electricity demand through nuclear power.[32] But instead they each produce around one-third of their electricity with renewables.

Many European countries are beginning to shut down nuclear plants in the face of fierce public opposition, France among these. Germany has taken an especially big role in the shuttering of nuclear power, planning to close all of its nuclear plants by 2022.[33] This may show to be a tragic mistake for the countries' energy mix. The idea around the globe for the reasoning behind shutting down nuclear plants has been to replace them immediately with solar and wind power. In some areas this makes sense in the face of the plummeting costs of these technologies, but in some other cases this can be disastrous. In many circumstances, a shuttered nuclear plant is replaced by a natural gas or coal plant, not by renewables. Germany is again a good example of this.

While Germany has been making massive strides to implement renewable energy (wind power became the country's second largest source of electricity generation in 2017, while the primary source remains to be coal),[34] the nuclear plants have been replaced mainly by two sources of power: wind and coal. Everyone (or at least, most of us) can agree that the closure of its nuclear plants for wind is either good or neutral from an environmental standpoint, but coal is a different story. This problem is only further amplified by the fact that German coal plants rely heavily on lignite (commonly referred to as "brown coal"), which is the dirtiest type of coal and has substantial air pollution implications. In 2018, Germany experienced nearly 66% more premature deaths from air pollution than France,[35] despite the population only being 27.6% greater than France's. As far as German natural gas, it is undoubtedly cleaner than coal but still increases ambient air pollution and GHG emissions when it replaces a nuclear plant. Additionally, Germany gets much of its natural gas from Russia, which has caused large tensions inside of Europe over concerns of energy security and has had a significant impact on US-German-Russian political relations. These tensions are yet again being seen with the fiasco regarding Nord Stream 2, a natural gas pipeline from Russia to Germany, which could result in sanctions against Germany by the US. In October of 2021, energy prices in the EU skyrocketed due to too little natural gas being provided by Russia. With the Russian invasion of Ukraine in 2022, the future of Nord Stream 2 is further being called into question.

It makes much more economic and environmental sense to first shut the coal plants, followed by natural gas and then finally shut the nuclear plants to be replaced by renewables. Many nuclear plants operate for up to 60 years, and many can have their lives extended to 80 years. When these plants are shut down after just 30 or 40 years of activity for political reasoning—as is the case with many around the world—there are massive potential economic benefits lost. Most nuclear plants were originally constructed with 40-year timetables in mind, and this life extension has caused many to sound the alarms of "unsafe lifetimes." In truth, this 40-year timetable was chosen simply as an ideal loan repayment plan timeframe and does not relate to the safety of aging reactors. As these reactors have aged, they have shown to be extremely durable and this is now why many are having their lifetimes extended.

Undoubtedly, the single biggest economic hindrance to current nuclear power technology is the cost of construction. Nuclear plants are notorious for being massive construction projects that are plagued by cost over-runs and construction-related delays.[36] This is by no means because of the amount of materials required for nuclear power, as is quite often brought up, but rather because of the complexity of what is being constructed and therefore the precision that is required, on top of all of the regulations that need met. In fact, when compared to other sources of energy, nuclear actually uses very few raw materials per unit of energy.

Many investors are simply not willing to put up with the upfront capital costs that nuclear requires compared to coal or natural gas. While the average nuclear plant in the US in 2016 cost $5,945/kW in Overnight Capital Cost (OCC, a simple metric to determine construction costs), the average Combustion Turbine (CT) natural gas plant cost only $1,101/kW in OCC.[37] Otherwise put, CT natural gas plant construction cost less than one-fifth that of nuclear. In this report, the OCC of other types of natural gas plants displayed a wide range of $678-1,342/kW. For consistency, this report also found that onshore wind had an OCC of $1,877/kW and PV solar had a range of $2,534-2,671/kW. The question you are likely considering at this moment is, then how is nuclear power economically competitive at all? The answer to this is twofold.

First, nuclear fuels are incredibly energy dense, and therefore, efficient. In fact, the uranium-235 used for nuclear fuel is over 160,000 times more energy dense than coal (which is different than the density comparison of coal to natural uranium as mentioned earlier, which is uranium-238).[38] While coal and natural gas may be cheap per volume, they require massive volumes. A typical nuclear reactor only requires refueling once a year, while coal and natural gas require a constant influx of resources. Because of this, nuclear power is extremely enticing when looking at it from a long-term perspective. On this metric, the tables are flipped. Over the lifespan of various power plants, nuclear frequently comes out on top as the most economically beneficial, whereas coal goes from (in almost every scenario) the cheapest form of energy to the most expensive. In a thermal coal power plant, the purchasing and handling of coal can make up 50-60% of the total operating cost of the facility.[39] (Natural gas also follows this trend—though not as drastically.) In a nuclear plant, fuel costs only account for 20% of the operational costs.[40] This long-term economic advantage is the primary reason why today France has such cheap and abundant electricity, while some of its neighbors are experiencing the opposite.

Secondly, nuclear power has an extremely high capacity factor. This is what helps to explain how it can be competitive with renewables. Whereas solar and wind (and hydropower to an extent) are heavily affected by intermittency driven by weather and climate—on a daily, seasonal, and yearly basis—nuclear is not. Let us compare a wind farm to a nuclear plant: Say your options are an offshore wind farm constructed for $5 billion, and a nuclear plant for $10 billion. The wind farm will only produce energy 45% of the time, but the nuclear plant will be producing power 90% of the time. Ignoring various complex aspects (such as interest, fuel cost, maintenance, and employment, all of which are significant), it will take the nuclear plant the same time pay off its loan debt as the wind farm, despite costing twice as much. After this loan is then paid off, the nuclear plant would become twice as profitable as the wind farm. Due to this high capacity factor, nuclear power also produces some of the lowest GHG emissions—over the lifetime of the plant—compared to other sources (see figure 7.4). This is to say nothing of how the wind turbines will need replaced every 30 years, while the nuclear plant will remain operating continuously over 60+ years.

Based on these factors, the only serious long-term competitors are hydroelectric and geothermal power, which have limited availability. For all of these reasons, it makes even less sense to shut down nuclear facilities before they reach the end of their life because they may have not even reached their full economic potential yet. Renewables will remain a strong competitor for the reason that they require no fuel distributed to them, but it will be massively difficult for renewables to replace the baseload power supply from nuclear. Also because of this, falling costs for these sources have an even more drastic effect, and will cause them to become cost competitive at a faster rate than other technologies (since manufacturing makes up a greater proportion of total costs).

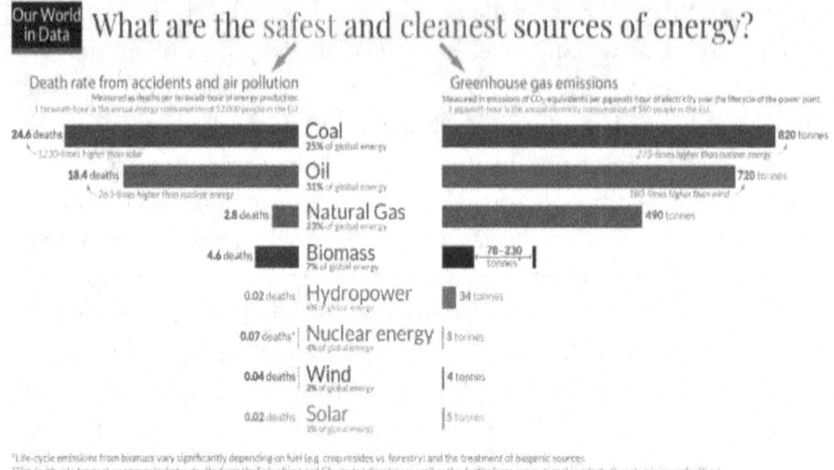

Figure 7.4: What are the safest and cleanest sources of energy?
Source: Ritchie, Hannah. "Nuclear Energy." *Our World in Data*, ourworldindata.org/nuclear-energy?country=#how-many-people-has-nuclear-energy-saved.

While construction costs plague nuclear power in the US, this is not the case in every country. The biggest issue with nuclear power plant construction in the US is that each plant is uniquely designed. This means minimal standardization, resulting in higher engineering costs and longer construction times. Countries such as South Korea and China have taken a different approach by standardizing their nuclear constructions to much more of an extent than is seen in the US. In these countries, a standardized plant is replicated over and over and is adjusted as needed depending on factors in the region. This top-down approach allows engineers and construction workers on each project to easily transfer their skills to a new project, reducing time and therefore costs for both design and construction. China is currently in the midst of the largest boom in nuclear power that has ever been seen, the growth rate of which is simply unprecedented. In just 15 years (2004-2019), China's nuclear power capacity increased nearly seven-fold.[41] Unfortunately (or fortunately, if you are against nuclear energy), this becomes very difficult to incorporate in the US due to each state having different regulations and a weaker central government.

Nuclear plants are still struggling to be competitive in US markets against natural gas. One of the major factors that could make already operating nuclear plants once-again competitive would be some form of carbon pricing (most likely through a carbon tax) that has been gaining increased traction. As it would make natural gas plants cost more through their carbon emissions, nuclear could make for an obvious choice depending on the specifics of the policy. In all reality, modern nuclear power is unlikely to hold out much longer in US markets but may still be viable in countries where natural gas is less of a competitor. Additionally, next-gen nuclear technologies could radically change this projection.

Summary of Modern Nuclear Energy Technology

Modern nuclear energy is not perfect. Primarily because of the problems it faces with regards to engineering and construction. It does, however, create magnitudes of less waste than other energies and is the only energy source that takes complete control of all its own waste The fact that nuclear power plants are incredibly expensive to construct causes many problems, again, to under-developed countries in particular where the financial incentives are non-existent.

It is this high cost budget, however, that is what the main focus of the next advancement in reactor designs will try to address. Research into these reactors has been looking extensively into the possibility of standardizing nuclear facilities so that they're all the same or very similar (to a much further extent than is even seen in South Korea and China). Some have even proposed that there are very realistic ways of man-ufacturing compact, lower power reactors on an assembly line-type con-figuration. This would drive the cost of nuclear fission construction and engineering down to fractions of what it currently is. There are obvious concerns here with proliferation and domestic terrorism (which are dis-cussed later in this section), but it is something that is showing to be a promising technology.

In the absence of new nuclear technologies, conventional nuclear power will still almost certainly need to play a large part of grid decar-bonization. Even while solar and wind go through massive booms in use, baseload power providers such as nuclear and hydro are going to be essential. The University of California Berkley's Goldman School of Public Policy's *2035 Report* looked at the practicality of having an electric grid that was 90% reliant on clean energy (mostly from wind and solar) by 2035. This report takes on the scenario of huge policy changes and cooperation in the energy field (whereas most estimates for this go based off more conservative estimates) that push the US towards this goal. Even in this most-optimistic scenario that projects massive increases in solar and wind, "retaining existing hydropower capacity and nuclear power capacity (after accounting for planned retirements) and about half of existing fossil fuel capacity, combined with 150 GW of new 4-hour battery storage, is sufficient to meet US electricity demand with a 90% clean grid in 2035…During normal periods of generation and demand, wind, solar, and batteries provide 70% of total annual generation, while hydropower and nuclear provide 20%." [42]

However, this estimate assumes that nuclear power is neither shut down more than already planned or expanded. As we have seen, there is a global trend in the US and across Europe to decommission nuclear plants. In 2019, 5.5 GW of additional nuclear capacity were connected to the global grid, but 9.4 GW were permanently shut down.[43] In the SDS, global nuclear capacity rises to 601 GW by 2040, but projected trends have nuclear power rising to just 456 GW—only 3% higher than the capacity was in 2020.[44] The SDS continues to state—in a subsection of recommended actions titled "Reduce nuclear policy uncertainty and recognise of the value of nuclear power in present and future low-carbon energy systems"—that "Nuclear policy uncertainty in a number of countries is preventing the nuclear industry from making investment decisions. This is partly the result of inconsistencies between stated policy goals – such as climate change mitigation – and policy actions. Forthright recognition by governments and international organizations of the value of nuclear energy's attributes and its contribution to decarbonizing the world's energy systems would encourage policy makers to explicitly include nuclear in their long-term energy plans and [nationally determined contributions] under the Paris Agreement".[45] It may sound somewhat dull to most people, but this is an example of strong language that the IEA and other UN organizations seldom use.

Alternatively, another problem that pro-nuclear advocates often leave out is that modern nuclear reactors take a long time to design and construct. Many of these plants can take well over ten years to be ready for operation, so while the potential for growth is massive, doing it on a timescale that is needed for climate change is very difficult. This is part of the reason why our solutions must include renewables plus nuclear, as they can help each other to grow sustainably and responsibly but also at a fast enough rate to combat climate change.

Overall, nuclear fission is not the perfect energy source, but it is extremely evident that it could (and will need to) play a very significant part as a key steppingstone between fossil fuels and a truly 100% clean world for the future. In truth, nuclear power may already play a bigger part in your life than you realize because we would not have access to much of our medical procedures (such as x-rays) without nuclear energy. And regardless of the medical improvements made with the help of advancing nuclear power that has so drastically revolutionized medicine in the developed world, around two-thirds of the global population already lives in countries that utilize nuclear power.

Nuclear fission should not, however, be the final end-all goal for electricity generation. The real ideal form of energy generation, as discussed later this section, is nuclear fusion. Nuclear fusion has the potential to drastically change the way that humans consume energy, and as far as we can tell, very well may be the perfect energy source. Fourth generation nuclear fission power will also be essential for providing the world with clean energy in the decades to come. However, before discussing those, it is worth taking time to review previous nuclear disasters.

Chapter 8

The Legacy of Incompetence
Nuclear Disasters

Nuclear energy used to be one of the most popular forms of energy generation. Governments saw great potential in reaping the energy of the atom and invested great deals of time and money into research and development. In fact, the world's first commercial nuclear power plant was commissioned by the Soviet Union in 1954, just nine years after the invention of the atomic bomb.[1] The US and Soviet Union were both very excited about the possible prospects of nuclear power and therefore expedited the development and rollout of nuclear reactors.

In the United States, the incident at the Three Mile Island (TMI) nuclear plant—located just south of Harrisburg, PA—in 1979 was when most Americans began to question nuclear energy. These worries were only then spurred on by the later incident at Chernobyl. Nuclear power slowly started making a comeback once the horrors of these events began to fade from the public conscious, but ever since the disaster at Fukushima, nuclear power has been on a steep decline in the US, Europe, and Japan. Some countries (such as China) have still been expanding nuclear at a high rate, but these disasters have almost certainly ensured that nuclear power will not grow much over the next several decades in Western Democracies.

These disasters, however, are often given a bad rap. In order to understand why nuclear power is one of the safest forms of energy production, a clear explanation of each of these disasters (and how they are blown way out of proportion by media outlets and other factors) is key. While there have been other incidents at nuclear plants, this chapter will focus on the big three: TMI, Chernobyl, and Fukushima. It is also important to remember that when there are explosions at nuclear plants, *they are not nuclear explosions*. A whole city is not instantly vaporized even in the most unlikely worst-case scenario.

Three Mile Island

The incident at TMI was doomed for bad press before it even started. Although TMI was the first major nuclear power incident (with the exception of the events at Windscale in 1957), environmentalists and fossil fuel interests (in their collaborative "war against nuclear") were becoming increasingly vocal with worries regarding nuclear power, and potential issues with nuclear power were starting to become commonly and publicly discussed. On March 16th, 1979, *The China Syndrome*—a successful thriller film whose plot takes place around the safety concerns of an unsafe nuclear plant, possibly causing major environmental and human health damage—was released, grossing over $4 million on its opening weekend. Just twelve days later, on March 28th, the partial meltdown of TMI's second reactor (hereafter referred to as TMI-2) occurred. The horribly bad and overexaggerated press was, at least in part, due to the stupendously unfortunate timing of *The China Syndrome*.

The problems at TMI began when there was a cooling malfunction in TMI-2. Several things were tested to restore full cooling, but due to inadequate training, faulty equipment, and poor plant designing, the reactor core became too hot, and some of the fuel (approximately one-third) melted, leaking radioactive materials into the water being used for cooling. Although TMI-2's reactor core was severely damaged, the reactor housing did what it was meant to do, which is contain radiation and waste.[2,3,4]

There was, of course, a small amount of radiation leaked from the plant at the end of this event. This happened due to some amounts of leakage from compressors that were used to move gases into disposal tanks. The amount leaked was miniscule. There are no three-eyed fish—as *The Simpsons* loved to depict, the species hilariously dubbed "Blinky"— that result(ed) from nuclear radiation.

There is extensive evidence of almost no harmful emissions released to the surrounding population or environment coming from this release of radiation. In fact, the leak of radiation experienced from TMI only gave, on average, people living within a 10 mile radius of the plant a radiation dose equivalent to that of an x-ray, which in turn is only equal to about 1-2.5% of the average background radiation received by a US citizen each year.[5,6] As mentioned before, we are exposed to radiation constantly; it is just a matter of putting the risks in context. This however did not matter in the eyes of many. The media quickly blew this incident out of proportion, and ever since then it is remembered as a "great nuclear disaster" even though in reality, it was at the worst a close call. This is not to say there was no significance about what had occurred that day, but the incident is almost always blown far out of proportion.* TMI was not a horrific disaster, but just a mere accident that was taken out of context. It is common to hear about the health effects and increased cancer rates over time around TMI. However, these frequent claims are not backed up by evidence anywhere near the equivalence of studies showing the opposite:

> "A team of researchers from the Graduate School of Public Health at the University of Pittsburgh, headed by Evelyn O. Talbott, reported in 2002 that they found no increase in rates of "radiosensitive" cancer mortality attributable to Three Mile Island among a cohort of 32,115 people who lived within a five-mile radius of the plant between 1979 and 1982. The study included members of the cohort through 1998, a period long enough for slow-developing cancers to show up. The richness of the data used in this project made it the gold standard of Three Mile Island epidemiology." [7]

In reality, the only true nuclear disasters that had an actual impact on life were what occurred at Chernobyl and (debatably) Fukushima.

Chernobyl

On April 26, 1986, it was the Soviets' turn to be reminded of the potentially devastating power of the atom. This was the date of the fateful accident at the Chernobyl Nuclear Power Plant. During a training exercise at the power plant, a malfunction in equipment and numerous

---

* This exaggeration will become a recurring theme throughout this chapter, but it is important to keep in mind that the initial reactions are not unjustified. During an event, like with the Covid-19 pandemic, you don't know how bad it could be at first, so it is often better to overreact than underreact. The important part is to put it back into context when the scientific data comes out and inform the public accordingly and do your best to not assume anything in the moment.

operator errors as well as a faulty reactor design caused a damaging power surge which hindered operators' ability to insert control rods, and the chain reaction could not be stopped. This generated large amounts of steam and then enough pressure to partially lift the 1,000-ton cover to the reactor. As steam was generated and became pressurized under the intense heat, it eventually resulted in an explosion. Following the steam explosion, a second explosion (likely due to the formation of hydrogen in the reactor) occurred. The containment shell of the reactor was compromised from these, and as raging fires burned, massive amounts of radiation continued to leak into the atmosphere. It has been estimated that over the course of the following ten days, half of Chernobyl's radioactive particles (particularly two called Iodine-131 and Caesium-137) and at least 5% of the remaining radioactive material were released into the atmosphere.[8,9]

The large majority of this quickly fell back to the ground, and most of it stayed within a 30 km radius around the plant. The radiation that stayed in the air for an extended period was carried by wind through Ukraine and into Belarus and modern-day Russia. Belarus in particular was hit with the majority of the radioactive fallout.[10] Although early panicking predictions estimated that the death toll could be in the tens or even hundreds of thousands, an IAEA report on the aftermath of the incident in 2005 estimates that about 4,000 eventual deaths could be expected from the radiation in the future, with around 50 deaths being directly attributable to the disaster. In the total population sample, this 4,000 is equivalent to about 3% of the expected typical cancer rates, making this very difficult to even identify.[11]

Chernobyl however was not caused by the "unsafe nature of nuclear power," but was caused by a combination of poor plant design by the USSR and inadequate training and knowledge by the staff. For example, nuclear reactors today (even the older models still used in many places) have containment structures around the reactor made of 3-5 foot-thick, steel reinforced, concrete walls, which were built to withstand a direct impact of an airliner even before 9/11.[12] At Chernobyl however, the RBMK reactor designs used had essentially no sort of containment building.[13] There was hardly any of this extra protective casing, meaning that the explosions easily blew a hole in the structure, releasing most of its hazards. Since the RBMK reactor design did not have a containment structure, not only was the radiation leaked, but the subsequent fire was provided oxygen to burn for an extended time period, all the while lifting radioactive material high into the air.

The impact that Chernobyl had on the nuclear power industry is profound. With an already worsening view among the public, Chernobyl caused major and permanent damage to the legitimacy of nuclear power. Although, as horrible as Chernobyl was, it may have helped nuclear power in some regards. At the time, the industry was lagging behind on safety features. While the RBMK reactor designs were not comparable to western designs, the incident still made the industry step up their game and improve the safety features in reactors. (Though this is a double-edged sword because while these safety features were great for the public, the rising costs and regulations associated with these adaptations are a primary driver of the rising costs of nuclear, making it increasingly less economic as time goes on and more regulations are incorporated.)

Chernobyl was the result of immense incompetence by a struggling, unstable totalitarian regime. This type of incident at this scale was incredibly unlikely to happen in other countries and is impossible today. The legacy of one nation's incompetence has plagued the nuclear power industry for decades and will continue to do so for years to come. In David Attenborough's 2020 film, *A Life On Our Planet*, both the beginning and end are filmed inside of Chernobyl's "uninhabitable" zone, capturing fascinating film in the process. Intriguingly, Attenborough does not bother to wear protection from the radiation—presumably because of his old age and his not being concerned about long-lasting effects in addition to the fact that radiation levels at Chernobyl are significantly lower today.

Fukushima Daiichi

Today's reactors are exponentially safer than those from the Cold War era. And although it was one of the largest nuclear disasters ever, the success of these new safety measures can be observed during the Fukushima Disaster in 2011.

On March 11, 2011, the largest earthquake ever recorded in Japan hit. This 9.0 magnitude earthquake off the Japanese coastline not only leveled vast parts of coastal cities near the epicenter, but also caused a massive tsunami that may have peaked as high as 133 ft. (40.5 m), traveling up to 6 miles inland at some points.[14] This tsunami rushed into Japan and devastated the coastline of Okuma, Fukushima, and other nearby provinces. The nuclear power plant Fukushima Daiichi had been built on the shoreline right where the tsunami and earthquake hit. The World Health Organization (WHO) reported that as a result the plant "lost its core cooling capacity which caused severe damage to the reactor's core and led to a nuclear accident rated as Level 7 on the International Nuclear Events Scale (INES). Substantial amounts of radioactive materials (radionuclides) were released into the environment following explosions at the [Fukushima Daiichi nuclear power station] on March 12, 14 and 15." [15] A level 7 threat is the highest level and was the same level given to the Chernobyl incident. Despite this, what happened at the plant before, during, and after the tsunami was quite impressive and not what many think.

To start, the earthquake that started the tsunami was so powerful that it moved the main island of Japan by about 8 feet,[16] and some early calculations predicted it may have shortened the length of Earth days and shifted the planet's axis.[17] Just think about that for a second. When this earthquake hit, all of the nuclear reactors affected (including all reactors at a second nearby nuclear power plant in Fukushima—Fukushima Daini) automatically began to shut down as a safety measure. The earthquake tripped the safety mechanisms in the plant, triggering them to lower the control rods into the reactors. By the time the tsunami hit the plant, the fission reaction inside the cores had largely been stopped, and it was more a matter of cooling the reactors at this point. Since cooling started immediately after the control rods were inserted, this also helped to reduce the impact of the disaster.

The structure was not heavily damaged by the earthquake while much of the rest of the city was heavily damaged, but it was the tsunami that became the problem. As the 15 meter tall wave crashed into the facility, it flooded everything, damaging many electrical aspects and hindering safety systems. A large part of the problem was that the rushing water disabled back-up generators that were acting to help cool the reactors. When the tsunami rushed in, it disabled twelve of thirteen back-up generators and this caused three of the four reactors to reach unsafe temperatures.[18] Because of the inability to continue rapidly cooling the reactors, several situations developed and contributed to radiation release. This includes the plant being overrun with water, meltdowns, hydrogen explosions (which are caused when steam is split into hydrogen and oxygen under extreme meltdown conditions—since hydrogen is extremely explosive, a small spark can ignite it, causing an explosion),[19] and other factors. Despite this, largely due to the safety features implemented, large amounts of radiation and nuclear wastes were kept contained until response teams were able to shut it down and secure the situation, though obviously there were also major leaks of radionuclides.

What really went wrong, however, is that the backup generators were not positioned properly and the sea wall for the plant was not high enough. In fact, the company operating the reactors—TEPCO—was aware of the risk of a larger than previously expected tsunamis and ignored the option to raise its seawalls to withstand 15m tall waves. This is exactly the height that ended up hitting the plant and causing all of these problems. Additionally, there had been warnings to move the backup generators to a higher elevation incase of flooding, but these were also ignored. If these points had been addressed instead of ignored, we would not be talking about Fukushima right now. It was the result of corporate greed due to the fact that these maintenances would have required the shut down of the reactors for a short period of time. By delaying this maintenance for short-term profits, TEPCO paid a heavy price.

After the natural disasters had passed, more aftershocks from the earthquake continued to hit in the following months. Even though the plant was leaking radiation, the structural integrity of the complex was mainly intact. Over the next month there were three more earthquakes near the plant measuring in at 7.1, 7.1, and 6.3. On top of this, there were reportedly hundreds of smaller aftershocks.[20] Despite this massive amount of destruction, the plant stood strong, without further damage from any of these aftershocks. This was not because of luck but due to proper construction of the site, leading it to be extremely resilient to the forces of mother nature. Whereas over 930,000 buildings in the region were damaged or destroyed,[21] the power plant infrastructure was left standing largely structurally unharmed—but flooded. In total, the WHO reported that the natural disaster caused 15,899 deaths and 2,529 missing persons.[22] None of these were attributable to the nuclear incident, with the exception of one person working to fix the plant post disaster who passed away in 2018 from radiation caused lung cancer.[23]

In fact, even the workers who died inside of the power plant during this event died as a result of the tsunami and earthquake, not by radiation poisoning (although some were exposed to high levels of radiation). Water and air radioactivity did increase, but much of the radiation that escaped had a short half-life, meaning that after just the first couple weeks the amount of radiation had already drastically decreased.[24] The radiation released from Fukushima was undoubtedly significant, but not nearly as much as was released from Chernobyl, despite the fantastical display of force put upon the plant. The IAEA estimated the total radioactive release from Chernobyl to be at 14,000 PBq [25] (a unit commonly used to represent radiation) and Fukushima was, by one other estimate, just 340-800 PBq.[26]

Figure 8.1 is from an IAEA report comparing the estimated release of two radionuclides, Iodine-131 and Caesium-137, from Fukushima compared to Chernobyl from various publications.[27] Though more information is needed, and the full effect of this incident may not become fully apparent until years from now, these preliminary findings show that the radioactive release is significantly less than many originally interpreted. Additionally, Fukushima is positioned to become one of the most well studied accidents in human history, specifically by the IAEA.

Generally speaking, after ten half-life decay cycles, the concentration of radioactive materials start to become insignificant. (At this point there is just 1/1,000 the original amount left.) The aforementioned Iodine-131 has a half-life of just eight days. Caseium-137 has a half-life much longer (30 years), but the amount released from Fukushima was quite small compared to the amount that is already in the atmosphere and oceans from nuclear weapons testing during the Cold War.[28]

Japan continues to closely monitor food and water in the affected areas to ensure that what is consumed is not dangerously contaminated. Many citizens affected were immediately relocated following the incident to avoid possible radiation poisoning and such, but as research continues to show, there was little threat of radiation to the area where the disaster occurred. There is still a large area of land around the plant not deemed safe, but some whose homes lie outside this area are already

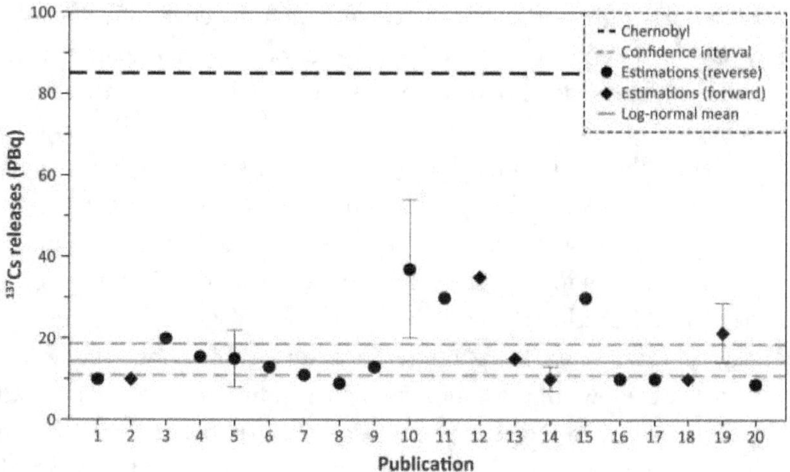

FIG. 1.4–4. Estimated atmospheric releases of ^{137}Cs (see Table 1.4–6).

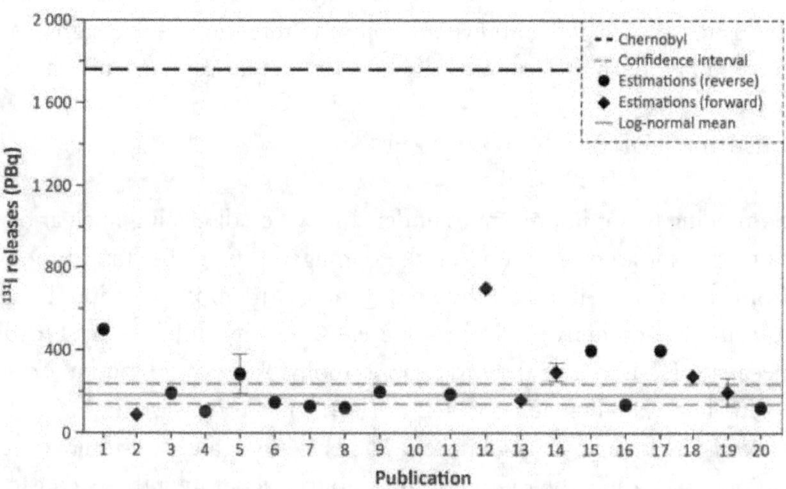

FIG. 1.4–5. Estimated atmospheric releases of ^{131}I (see Table 1.4–6).

Figure 8.1: Estimated atmospheric releases of I-131 & Cs-137

Source: INTERNATIONAL ATOMIC ENERGY AGENCY, The Fukushima Daiichi Accident, Technical Volume 1, IAEA, Vienna (2015). See citation "27" this chapter for more details. Taken with permission by the IAEA.

moving back. Furthermore, it is estimated that the effect of relocation may have had significant psychological and sociological impacts on the population. The sudden relocation of some 200,000 people always has significant impacts.[29]

Overall, the Fukushima disaster was a tragedy. And while this reminded the world that there can still be flaws and serious dangers of nuclear power, it also raised the perspective that modern plants remain relatively safe even in the absolute worst-case scenario. Although some areas around Fukushima might not be considered habitable for humans for several more years, the plant performed most of its job quite well.*

This also shows that as humans, we learn from our mistakes. Had the reactors been an RBMK reactor like at Chernobyl, the deaths from radiation could have been significant. Also, quite ironically, inside of the "uninhabitable" areas around Fukushima, wildlife is flourishing. Because of the lack of human presence in these areas wildlife presence seems to be stronger than ever, despite the radiation.[30] The same can be seen in Chernobyl, despite higher radiation levels than Fukushima.

Summary of Nuclear Disasters

Something that is important to understand when discussing nuclear disasters is that a large portion of the damage done to the human body from radiation is not actually fatal. One of the most prominent side effects of radioactive fallout is the cause of thyroid cancer. This is because of I-131. The thyroid extracts iodine from your blood in order to create certain hormones which will perform various tasks in the body. Since I-131 is chemically identical as other natural (and non-radioactive) isotopes of iodine, the thyroid is incapable of separating these, causing I-131 to concentrate in the thyroid, often resulting in cancer. This is why in the case of nuclear fallout you are instructed to take iodine pills, which will dilute the concentration of the radioactive iodine in your thyroid (if you live close enough to a nuclear power plant you may have been supplied or notified to have iodine pills in case of emergency. If you aren't sure if you should have iodine pills, you likely don't need

---

* Despite the radiation, an increasing amount of people continue to relocate back into the region. Most of the land is expected to be safe within several years, though more information is still needed. Large areas that were originally contaminated by radiation are already perfectly safe again.

them). Fortunately, there is a silver lining here. Thyroid cancer is one of the easiest types of cancer to treat, with a survival rate of over 99%. And if iodine pills are ingested, the likelihood of forming thyroid cancer is dramatically reduced. This is to say nothing of the fact that I-131 has a half-life of only eight days. After just a few months, I-131 levels in both Fukushima and Chernobyl had dropped significantly to the point where it was no longer of concern.

As for the remaining radiation, the risks also rapidly fall as time goes. Today, the radiation levels at both Chernobyl and Fukushima are drastically lower than at the onset of the incident. The biggest threat of these sites remains to be the ingestion of radioactive particles, as large amounts of the radiation remaining is not very harmful to people when they are exposed to it from outside of their bodies (alpha and beta radiation can be easily stopped from entering the body by a thin layer of clothing or even the layer of dead skin on the outside of your body). While gamma radiation and x-rays (which can penetrate through thicker layers of clothing and skin) from the surrounding environment remain a threat, the levels of this in both of these sites is rapidly decreasing as well. Today, large areas of these sites that are still labeled as "inhospitable" are, in fact, no more dangerous than high-radiation places around the globe that are inhabited by thousands of people—such as Ramsar, Iran, and the stunning black beaches in Brazil—which show no higher rates of cancer or radiation sickness than normal.

TMI, Chernobyl, and Fukushima Daiichi are just the most-well known nuclear incidents. There have been a number of other nuclear incidents—most notably the close call at the Windscale nuclear plant in England—and many estimates total well over 100, but most of these never gain much attention because they are, in the big picture, insignificant. Does this mean that nuclear power *is* dangerous? No. Does this mean that nuclear power *can be* dangerous? Certainly. But energy production is never risk free. Considering its track record, it is one of the safest forms of energy production that we have. If you take a high estimate of 5,000 people killed (including the projected 4,000 deaths from Chernobyl that are yet to be seen), worldwide, from commercial nuclear power over 65 years of commercial operations, it becomes easy to see how much worries about safety are blown out of proportion.

5,000 divided by 65 years gives us an average of 0.21 fatalities per day (f/d). Meanwhile, in just the US, 32,000 die from motor crashes each year—an average of 87.67 f/d.[31] Cigarette smoking in the US causes 480,000 deaths each year—1,315 f/d.[32] At the time of this writing, the Covid-19 pandemic was (as reported by the WHO) responsible for the deaths of 725,000 Americans—1,208 f/d. Globally, this figure is over 8,000 f/d (assuming the start of the pandemic at March 1st, 2020).

As will be discussed later in this book, air pollution—mainly from the burning of fossil fuels and biomass but also cigarette smoke, and various other environmental factors—is responsible for over 7 million deaths globally each year—19,178 f/d.[33]

Nuclear power is scary because of the fact that we can't directly see radiation, and it makes the populous uncomfortable. It is much easier for humans to feel comfortable with a problem that is known to cause serious health effects but easily identifiable (fossil fuels and their ensuing air pollution) than to deal with something whose effects are minor but isn't easily identifiable by the average person (nuclear radiation concerns). In his book, *How to Avoid a Climate Disaster*, Bill Gates sums up this relationship best: "We have a large and understandable incentive to stick with what we know, even if what we know is killing us." Radiation simply instills a fear of the unknown.

Chapter 9

Thirty More Years...
Nuclear Fusion, a Future Solution

Nuclear fusion is the other type of nuclear reaction. In fact, fusion is the exact opposite of fission. Whereas fission splits atoms to release energy, fusion forces two atoms together, which also releases energy (and much more of it, nuclear fusion is literally what powers the stars). Super condensed matter in the sun at its core creates heat at 27 million degrees Fahrenheit,[1] eventually creating enough heat and pressure to push atoms together. This forms the process of nuclear fusion, giving us the light and heat that provides our Earth with the conditions for life. Therefore, nuclear fusion in a way is the giver of life to the universe. Nuclear fusion for electricity generation on Earth could provide us with nearly infinite amounts of power that is cheaper than any other type of power generation, exceedingly clean, completely meltdown proof, all the while producing next to no nuclear waste. Even those who are against conventional nuclear (fission) power have a very hard time arguing against fusion energy. So why aren't there already fusion power plants all around the world yet?

To start, some background on nuclear fusion. As explained, nuclear fusion is much more powerful than fission. Although the first atomic bombs were fission-based, large-yield nuclear weapons ever since the 1950s have been fusion-based (commonly referred to as "Thermonuclear" Weapons). The world's first fusion-based weapon, "Ivy Mike," was detonated by the US in 1952 and had a blast yield roughly 700 times more powerful than the atomic bomb dropped on Hiroshima, Japan during World War II.[2] Farther down the road towards the end of the Cold War, the Soviet Union detonated a bomb, dubbed "Tsar Bomba" (Russian for "King of Bombs"), which was about five times stronger than that of Ivy Mike. This bomb was so powerful that the explosion flash could be seen from up to 1,000 km

(621 miles) away and it even shattered windows in towns as far as 900 km away.[3] The heat from this was so intense it could be felt from more than 270 km away.[4] That's comparable to detonating the bomb in New York City and shattering windows in Cincinnati. The atmospheric disturbance caused by the bomb circled around the Earth three times.[5] If that wasn't enough for you, the weapon was originally made to be twice as powerful as this. Before conducting the test, the Soviets decided to cut the yield in half to reduce the amount of radioactive fallout.[6] It is genuinely hard to comprehend the amount of energy we would have available to us by harnessing this source in a structured and contained way. We would easily be able to power everything in our modern world and more with the use of obtaining a power source so powerful. This is likely why nuclear fusion has become part of the world of Sci-Fi, showing up in movies like *Interstellar* and the *Back to the Future* series.

The main drawback to fusion power—and this is a big one—is that we can't quite do it. Not yet anyways. Since the invention of thermonuclear weapons, scientists around the world have been working hard at creating fusion energy and, until very recently, have been relatively unsuccessful. We have successfully been able to create semi-controlled fusion reactions where we can produce energy since the 1970s, but progress stagnated for a long period of time. With a renewed interest in the technology, extensive progress has been made this century. Still, two main problems with realistically creating fusion energy remain. First, we have to put more energy into these reactors than we get back out of them, as it takes energy to begin and sustain the reaction. And second, the fusion reaction is incredibly difficult to maintain in a controlled manner for an extended period of time.

Solving Fusion

"Q" is what some physicists use to determine the energy return from a system. If $Q=1$ then you received the same amount of energy from a reaction as you put into it. If $Q=2$, then you received twice as much energy as you put into the system. The goal should obviously be to get a Q ratio as high as possible. Fusion experiments in the past have been largely unsuccessful in this area. So far, the most successful fusion reactor has been the Joint European Torus (JET). JET generated just 16MW of power from 24MW, giving it a Q of 0.67.[7] It is also important to note that Q in this case is referring to the energy return from the fusion plasma, and this is not representative of the whole system. ITER's plasma may get a Q=10, but the site will likely only achieve Q=0.5 in total. This can largely be attributed to energy loss from turning the heat into electricity.

However, this might change very soon. As fusion became harder and harder, countries eventually came together to collaborate and share knowledge. The ultimate result of this was the International Thermonuclear Experimental Reactor (ITER, pronounced "eater"). ITER is a joint project involving China, the European Union, Switzerland, India, Japan, South Korea, Russia, and the United States (35 countries in total).[8] ITER is looking to change the long-standing history of failure in fusion energy. A large-scale fusion reactor, ITER is expected to produce its first plasma in late 2025. Once fired up to full working capacity, ITER is projected to grant a Q=10.[9] But even after ITER, this collaboration is far from done their work.

If ITER does indeed prove to be as successful as projected, then there are plans to begin construction of a new reactor called DEMO. ITER will not be used as an example of what fusion power plants should be, but instead will be used to understand the principles and extreme physics of fusion better and use this knowledge to improve upon the design for DEMO. Planning is already underway for DEMO, but this machine is still far away. DEMO is what is being proposed to be a demonstration of what a real nuclear fusion power plant will look like. Unfortunately, early estimates say that construction of this facility won't begin until the 2030s and operation in the 2040s, and it's very possible this is pushed further back. Although, if everything does go according to schedule, then it will still take several decades before we can have expected to implement this technology into our power grids for commercial use.

Although DEMO's concept is for relatively far in the future, the possibilities are encouraging. Early proposals have classified DEMO to have a power output range somewhere between a Q-value of 30-50,[10] though it is likely these numbers change drastically in the future depending on results obtained from ITER. Even if achieved at 30, this amount of energy is very significant. It has even been proposed that if fusion were effectively achieved to this, getting electricity for your home could be as simple as paying a small flat rate every month just like internet or cable TV. And this would be a very small price too—some predict that electricity will become even cheaper than water in this scenario.*

Nuclear fusion's other problem is that it is very difficult to keep the nuclear reaction consistent. This is quite a task and without a doubt the biggest technological hurdle that needs to be overcome. On the sun, there are massive amounts of gravity (and therefore pressure) and heat, which cause fusion to occur naturally, but here on Earth we are missing these factors. We are able to create heat, but to create the extreme pressures is a big challenge. However, it turns out that if you can go hot enough, then you do not need the immense pressures found in the sun.

As the atoms required for the fusion reaction become hotter and hotter, they move increasingly fast and become more apt to fusing together in a nuclear reaction. The problem here is that instead of heating these atoms to the temperature of the sun, they need to become even hotter due to the lack of pressure. In fact, ITER is set to produce temperatures up to 150 million degrees Celsius. That's ten times hotter than the center of the sun.[11] It is worth mentioning that there are some private fusion companies that are attempting to create fusion through pressure, but the primary method is through a tokomak device, which uses heat and magnetic forces for a fusion reaction and is the specific type of contraption for fusion that is main the focus of this chapter. While private interest is exciting, most of these companies are relatively new and do not seem quite as progressed as projects like ITER, though recent breakthroughs at MIT, Helion Energy, and others do show some promise.

(Additionally, private interests do not have a great history undertaking massively complex new engineering projects like this. His-

---

* This is almost certainly not going to happen. The concept of energy "Too Cheap to Meter" was also proposed with nuclear fission power in the early 20th century, and some have again taken to saying a fission comeback or widespread use of renewables could produce this. The reality is, these claims are dreams of a future, not a prediction of what will be.

torically, comparable feats have always been pioneered by government organizations and the technology is then later taken by private groups to be improved upon. The prime example of this is with regards to space. While NASA created the modern space industry, private companies such as SpaceX, Blue Origin, Virgin Group, Rocket Lab, and others are becoming increasingly dominant in the industry. The mixed economy that leverages the massive resources of the government and then later couples it with the competitive ingenuity of private business has proved well for these types of projects in the past since it is risky and difficult for a private business to create something of this scope while also having to worry about turning a profit. With industries such as space and nuclear fusion, there may be decades between having the idea and making it profitable, all while spending billions of dollars in research.)

When the gas inside the reactor reaches temperatures this blisteringly hot, it is turned into a plasma state. Because of the extreme temperatures, this plasma would instantly incinerate anything it came into direct contact with. This obviously causes a big problem containment wise. However, when a substance goes from a gas to a plasma, it becomes extremely susceptible to magnetic forces. So, what a tokomak achieves is being able to keep the plasma contained at the center of the reactor through the use of said magnetic forces.[12] Even still, all of these forces are exceedingly difficult to control in a regulated fashion. ITER has made extensive progress in this field, and when firing up the reactor they are expected to solve the final few problems, which will be applied to DEMO. Should this project fail to find a solution to fix these problems however, fusion energy may become abandoned since it is already long overdue. It may simply be seen as something that we, as a species, are not technologically capable of yet, though I find this doubtful.

Safety and Waste Concerns

The power of the atom has been feared ever since the invention of the atomic bomb. When we began to use nuclear power to give us our electricity, worries about what could happen if something went wrong were common. These fears then became reality with a string of disasters (see chapter 8), Chernobyl being particularly frightening. And with how much more powerful thermonuclear weapons are than conventional nuclear weapons, these worries can be amplified even further with fusion technology. However, these disasters should have no effect on the view of fusion energy because they simply don't apply to a tokomak-style fusion reactor.

As explained, fusion is the opposite of fission. And although we are able to make bombs from it, a power plant does not work like a bomb. What we can see with nuclear fusion reactors is that they cannot melt down like with a fission reactor. This doesn't mean they are less likely or more resistant to a disaster; it is *physically impossible*. Because a fission reaction has a chain reaction, the reaction can become unstable and grow out of control if not properly regulated, which can result in what we call "meltdowns," where the actual nuclear core melts. In a fusion reactor, the reactants are being combined together and are not involving other particles around them in the reaction. Therefore, no chain reaction is occurring, and as soon as there would be a malfunction, the reaction would cease to stop immediately. There isn't even a nuclear core like there is with fission. The closest comparison would be how a solar panel ceases to produce power when the sun goes away.

As stated by ITER, "A Fukushima-type nuclear accident is not possible in a tokamak fusion device. It is difficult enough to reach and maintain the precise conditions necessary for fusion—if any disturbance occurs, the plasma cools within seconds and the reaction stops. The quantity of fuel present in the vessel at any one time is enough for a few seconds only and there is no risk of a chain reaction."[13] It isn't even possible to manipulate this reactor to become a bomb if we wanted. In thermonuclear weapons, the fusion explosion is still initiated by an extremely specific chemical and/or fission nuclear explosion. This explosion in these bombs has to be so precise that even a fraction of a second difference in the timing of the explosives could result in the thermonuclear blast not happening at all—which has happened several times throughout history. This series of explosions leading to the full payload are so precise, the entire sequence of events occurs in about 600 billionths—0.0000006—of a second.[14] With all of these factors combined, it is simply not possible for there to be any type of explosion or meltdown inside a fusion reactor.

Just like fission, fusion power would produce no harmful or GHG emissions. Though the construction of these plants would be immense and would create a lot of emissions, the emission saving potential of these would handily outweigh their construction emissions. Nuclear waste is also a huge benefit of nuclear fusion. The main byproduct produced by a fusion reaction at a plant like ITER would be helium gas.[15] Helium is an inert gas and has no greenhouse effect potential. Furthermore, helium is used in many industries—most notably in the manufacturing industry. Because of its usefulness and the fact that the world is rapidly running out of helium (which makes the using of the substance for party balloons absurd),[16] the helium produced from a reaction could even be sold from the facility for extra profit. It is important to note that fusion will produce some radioactive waste, but it is not nearly as radioactive as the byproducts from fission and is very short lived. ITER claims the components of the reactor can be recycled after 100 years.

The waste created other than helium in some tokomak devices would be a substance called tritium. Tritium is a radioactive isotope of hydrogen but is far less radioactive than the plutonium or other radionuclides developed as a byproduct through uranium fission. In general, tritium isn't even considered to be radioactive enough to be dangerous to human health unless exposed to in huge quantities or ingested.[17] Tritium also has a significantly shorter half-life than current nuclear waste. Whereas various fission waste products can have a half-life of hundreds of years, the small amounts of tritium that is both used for and created from fusion has a half-life of only 12.3 years. This means that the tritium loses its radioactivity very quickly (compared to fission waste), and we can take it out of storage and into any normal sort of waste storage after it's decayed.

In truth, none of this really matters because although tritium is a byproduct of fusion, it is also a starting component. Although tritium is almost non-existent naturally due to its short half-life, it can be created inside of the fusion reactor with a lithium breeding blanket.[18] In essence, this allows us to create the fuel for the reactor inside the reactor itself, allowing for large amounts of tritium to be created and used in a controlled and safe manner. This is also where many of the "energy too cheap to meter" claims of fusion power come from. This breeding process combined with the fact that there is already much less, and not as radioactive waste compared to fission reactions, shows that waste management for a deuterium-tritium fusion reaction (as ITER is) will prove to be an easy problem to handle and keep in check. This issue is even further compounded by the fact that, as discussed earlier this book, nuclear fission waste itself is not nearly as much of a problem as it is made out to be.

Other Advantages

Possibly the single largest advantage of nuclear fusion to compete with renewables and fossil fuels is the length of time that these reactions could be feasibly sustained on Earth. As far as fossil fuels, "There are an estimated 1.1 trillion [metric tons] of proven coal reserves worldwide. This means that there is enough coal to last us around 150 years at current rates of production. In contrast, proven oil and gas reserves are equivalent to around 50 and 52 years at current production levels."[19] * There is enough uranium to run the world at current production rates for over 200 years and likely significantly more than that if we utilize the uranium in more efficient manners than we currently do,[20] and enough thorium to last thousands (potentially tens of thousands) of years. (More on thorium next chapter.) With nuclear fusion, it has been estimated that there is enough easily accessible fuel available on Earth to power the whole planet for even longer than this, possibly for hundreds of thousands or millions of years. This is largely due to two factors.

Nuclear fusion relies on just two distinct materials for fuel. These are both types of hydrogen, Deuterium (H-2) and Tritium (H-3). Like tritium, the deuterium used in nuclear fusion is an isotope of hydrogen, making it extremely common in sea water. Though tritium is not common naturally (due to its radioactive nature), we can fabricate tritium outside of the natural world due to this breeding blanket, and so the only constraints are resources for the blanket and the supply of deuterium.[21] Of this, our supply of lithium would likely become more of a concern before we run short of deuterium (of which could take

---

* Note that environmental, nuclear power, renewable power, governmental and academic individuals and organizations have long predicted global shortages of fossil fuels but have been proven wrong over and over.

millions of years).

Secondly, since the fusion reaction produces magnitudes more energy than traditional fossil fuels, a small amount of this said abundant resource goes a long way. So much so that the amount of deuterium present in one liter of water can in theory produce as much energy as the combustion of 300 liters of oil.[22] For these reasons, fusion is sometimes considered a renewable resource.

With manned missions to Mars and the Moon already planned for within the next couple decades by NASA and SpaceX, it is extremely likely that by the time we would even begin to make a sizable dent in the amount of fusion fuel on the planet, we will have already begun to colonize (or have fully colonized) Mars, the Moon, and other planets around the solar system as well, being able to gather the necessary resources from those places. Speaking of the Moon, it has quite the potential to provide us with the ultimate form of nuclear fusion. The Moon could have large amounts of a substance called Helium-3, where mining operations could hypothetically become an extremely lucrative business.[23] The potential advantages to Helium-3 fusion would be that it is simply a much more controllable fuel for the nuclear fusion process. This would give fusion a further economic advantage over other forms of energy, but this endeavor is still far away.

Limits of Fusion Power

A significant problem with nuclear fusion is that it is extremely expensive. ITER is one of the most expensive research projects ever conducted in human history, and the costs are only going to increase until the technology is ready for industry. ITER is one of the most expensive sites in the world, with a reported cost estimate of $22 billion, though the US Department of Energy (DOE) has estimated this cost to be closer to $65 billion.[24] Commercial nuclear fusion reactor sites in practice would not cost nearly this much individually, but there is no doubt that they will still be very financially demanding, perhaps on a scale similar to what is seen with current nuclear fission costs in the United States. If fusion energy becomes commercially viable in the future, then this will be its highest hurdle. Building fusion power plants in the future (assuming the absence of major breakthroughs outside of

ITER) will be large, complex, and extremely expensive projects.

This means that in the absence of long-distance transmission cables, these power plants will never be as viable to power small, rural areas as other decentralized sources such as wind and solar. This also reasons that poorer, and less developed countries will be left behind in this energy struggle, and as they go through their own industrial revolutions, they will likely turn primarily to fossil fuels to meet their needs. For this reason, it will be the more developed countries of the world like Russia, China, India and Western Society that can really take advantage of this opportunity. Although it is unfortunate that many countries might not be able to use this technology (at least not until far in the future), this itself is not an absolute tragedy for the climate.

In 2018, 55% of the world's CO_2 emissions came from just four countries: China, Russia, India, and the US (with China and the US contributing 28% and 15%, respectively).[25] If you include other developed countries such as Japan, Australia, and European nations, only about 25% of the world's carbon emissions come from undeveloped or developing countries.[26] While this means that fusion can help significantly for most of the world's current energy needs, this does not account for how the energy needs of the world will shift towards developing countries—specifically in Sub-Saharan Africa. It is estimated that by the end of this century, 35% of the world's population could be living in Sub-Saharan Africa, with an additional 8.4% from Northern Africa and Western Asia.[27]

This dramatic shift will have quite an impact on both climate change and the economic structure of the world. If these countries would indeed choose to turn to fossil fuels, then the impacts on climate change efforts would be greatly affected. Thus, for a fusion-powered world to succeed, it will be imperative that more developed countries fund these projects in smaller developing countries and incentivize them to take initiative into their own hands when the time is right. This could not only add hope for a cleaner world but could be economically bene-ficial for all parties if done correctly. Fusion plants would have very comparable economic benefits as discussed with fission, as they will re-quire a large quantity of highly skilled, high wage workers, not to mention the public health benefits from removing fossil fuels and miti-gating electronic waste from certain renewable energies.

Summary of Nuclear Fusion

It is extremely important to note that this is a type of power plant in the most normal way we imagine it. Unlike solar and wind power, little infrastructure would actually need changed or added to implement these aside from the power plants themselves. Additionally, unlike solar or wind, this is a consistent flow of energy that we could adjust as needed. This wouldn't rely on the sun to be shining or the wind to be blowing for us to produce power. A step further, although it is likely that fusion facilities would be massive, they would likely not even be close to the amount of space that we would occupy from trying to power the world exclusively from renewable sources. To produce 500 megawatts of energy, an average wind farm would need to cover about 50 square miles (though in practice much of this land is not actively being occupied by the turbines, as was discussed in chapter 3).[28] ITER is projected to produce the same amount of energy in 180 hectares (~¾ of a square mile).[29]

Nuclear fusion could be the energy source to power human civilization for forever, but it is still going to take a while to get here. We need to reduce the emissions we are producing now, and nuclear fusion is already behind the curve. We may be able to utilize fusion in the second half of the 21st century, but until then we desperately need a transitioning source of electric generation. Modern nuclear fission is an option, but increasingly unpopular and expensive. In the absence of practical storage solutions for renewable power, the only option left in some instances may be natural gas—which emits half the GHG emissions as coal, but is still far from a true attempt to get off fossil fuels. Even though it's estimated to be here by the 2040s or '50s, fusion energy has a long-standing history of being delayed. Scientists have been saying "30 more years" since the 1950s, and we cannot sit back and wait forever for this technology to get here. There is a running joke about the industry that goes "Nuclear fusion is only 30 years away—and always will be."

So, we need a source of power that can act as a transition off fossil fuels and towards the clean utopia that comes with nuclear fusion and is either available now or will be available soon (within the next 5-10 years). This is the role that next-generation nuclear power, coupled with various renewable technologies can play.

Chapter 10

The Solution We Need

Next-Gen. Nuclear Fission, a Near-Future Solution

As mentioned earlier in this section, modern nuclear energy has potential but is far from a perfect technology. With its construction issues and increasing regulation by international and state organizations, this stagnating growth trend seems likely to keep up in the future, which will prove conventional nuclear power generation difficult to implement—particularly in regions that are not fully economically developed or are socially/politically unstable.

In the hindsight of nuclear fusion "always being 30 years away" and with the potential prohibitive upfront costs, nuclear fusion is almost certain to not be a realistic option for utility-scale power generation until, at the earliest, the second half of the 21st century. Because of the fact that major carbon-free sources of energy need developed and implemented yesterday, there is simply not time to wait for nuclear fusion.

These discrepancies between the past, present, and the distant future are what cause many to dismiss nuclear technology altogether. The fatal flaw in this ideology, however, is that this does not consider the reality that is the *near future* of nuclear technology. The near future of nuclear power is fourth generation (here-forth referred to as gen. IV) nuclear fission. The critical distinction between near future and distant future nuclear technologies is that the near future gen. IV reactors are not simply theoretical, as is still largely the case with nuclear fusion. (Again, ITER and other nuclear fusion technologies may change this perspective in the coming years.) Gen. IV reactors, in many cases, are a reality that is simply waiting to be accepted and implemented. As we will see, in some cases, gen. IV reactor technology was developed as early as the 1960s and was sidelined for other technologies that would better fit military (instead of civilian) purposes.

Gen. IV nuclear technology has massive potential to completely change the way that nuclear power is viewed. There are four distinct areas of focus of gen. IV nuclear power that are internationally agreed on. Sustainability, economics, safety and reliability, and proliferation resistance and physical protection.[1]

The discussion for a fundamentally different type of utility-scale nuclear energy has been in talks since early in nuclear powers development. Many who worked on developing early nuclear power technology were aware of the possible risks that our now modern nuclear power may pose. Nuclear waste, meltdowns, difficulties in cost, these were all aspects that were well discussed and worried about early in nuclear powers design.

Due to efforts to create nuclear weapons in the face of the Cold War, what we would largely consider gen. IV reactors today were sidelined for other types of reactors—specifically WCRs and breeder reactors—that were better at producing weapons grade materials (specifically plutonium) through the nuclear reaction. Plutonium is especially useful for nuclear weapons, and since it is only found in trace quantities, nuclear reactors were designed to produce weapons grade materials just as much as provide energy. Truth be told, the power generation was more of an afterthought.

Just so there is no confusion, keep in mind that in the US there has only ever existed one reactor that ever actually operated as dual purpose for providing weapons grade plutonium *and* producing electricity—a reactor dubbed "N" located in Hanford, Washington state.[2] The key takeaway, however, is that although nuclear plants were seldom actually used for dual purposes outside of the Soviet Union, the technology we have today derives from this concept nonetheless. With limited budgets, angry military generals, and a Cold War looming overhead, the decision to research on one type of nuclear technology that would have the ability to do both of these functions was made. While this served well enough for several decades, there were many researchers who knew the safety risks and plethora of downsides that came with what we today consider conventional nuclear fission. This decision ultimately was therefore responsible for the nuclear disasters at Three Mile Island (TMI), Chernobyl, and Fukushima, as well as the gradual decline of the nuclear power industry. In some ways, gen. III reactors (also commonly referred to as "Advanced Nuclear" reactors) are a newer creation than what will eventually become the gen. IV reactors of tomorrow. Before we explore how and why gen. IV reactors will completely change the way the game is played, it will be helpful to understand the basics of previous generations of nuclear technology.

A Brief History of Early Nuclear Power (Gen. I-III)

What is typically referred to as gen. I nuclear power was not widely implemented or designed for energy use at commercial scale (though there was some commercial adoption). These mainly operated as proofs of concept for nuclear energy at national laboratories. All of the gen. I commercial reactors have since been updated to newer reactors, or (as is the case with most) closed. Because of these factors, gen. I nuclear was obsolete once the technology had been improved upon, and by the late '60s gen. II nuclear energy began to become much more prevalent. It was in this stage (gen. I) that the decisions to not focus on more advanced reactors that would be better suited for civilian applications were made. The advanced technology that we are exploring again today never took off in the nuclear industry because it was never able to get out of the labs. As we will see later this chapter, it was around just long enough to prove its concepts, but never received the funding to get onto real-world applications.

Gen. II reactors took the information learned from gen. I and turned it into real power plants. Various types of plants resulted from this process. This includes familiar technologies such as PWRs, BWRs and LWRs, as well as older designs such as the **RBMK** reactors used by the Soviet Union. Although, as all things, this technology was improved upon over the decades to come, the nuclear power industry would not see a dramatic shift in nuclear technology until the 1990s. Gen. II technology was designed from gen. I technology with the mindset (one might even go as far to call it an ideology) of propelling man into the atomic age. In the '50s, the "power of the atom" was being touted for the potential to one day completely revolutionize society. Not just because of electricity production, but claims were abundant that nuclear power and the harnessing of radiation would revolutionize medical science, agriculture, and industry. (Of course, nuclear did revolutionize power production and the medical field. It had some impacts on agriculture and industry but not to the amount that was propagated.)

At one point, even radioactive cosmetics and toothpaste were marketed in Germany.[3] Because radioactive substances glow in the dark, radioactive watch paint during the first few decades of last century was used so that people could tell the time at night, which gave some of the women tasked with painting the watches cancer, leading to the infamous '*Radium Girls*' incidents. Propaganda depicting the coming of the atomic age were prevalent in the US with films such as Walt Disney's *Our Friend The Atom* in 1957 and encouraged by former President Dwight Eisenhower's *Atoms for Peace* program.[4] In 1953, Eisenhower asked the UN to create an international organization for the promotion of nuclear power.[5] This would, of course, eventually become the International Atomic Energy Agency (IAEA).

In the face of all of these promises from the first half of the 20th century and the 1950s, nuclear energy was propelled forward at light-speed and quickly rushed into development in every possible circumstance. The most prominent example of this is obviously what would become gen. II nuclear power plants. While this rush caused much positive R&D in the nuclear field, the time constraints caused certain liberties to be taken. This was gen. II nuclear technologies' kryptonite. After the incidents at TMI and Chernobyl, the public's infatuation with nuclear quickly faded and turned against the technology. In recent history, the Fukushima nuclear incident once again gave the nuclear power industry trouble. Interestingly, all of these reactors were gen. II reactors. (Chernobyl had a mix of gen. I & II reactors, but the reactor that exploded was a gen. II RBMK reactor design).

These incidents (mainly TMI and Chernobyl) were what pushed forward the development of gen. II+ and III reactors. Gen. II+ designs focused heavily on refurbishing existing reactors for safety, and most reactors today lay somewhere between II and III.

Gen. III technology is not inherently different in most ways than gen. II technology. The premise for this was mostly to increase safety and improve constructability. This includes passive safety systems (meaning the reactor can stop itself from a meltdown without human interference) and standardization of plant designs.[6] (There is also what some consider gen. III$^+$, which takes this another step forward, but we will refer to all of them simply as gen. III).

Unfortunately, while gen. III is considered the "modern" nuclear technology, the reality is that most of the world's nuclear reactors are still gen. II designs, leaving nuclear power largely in the past. Of the United States 96 operating reactors,[7] none are gen. III (though there are a couple gen. III reactors under construction). This is a critically important point. Anti-nuclear activists commonly use the argument that our nuclear plants are outdated and not safe, while advocates for nuclear power argue that while nuclear energy used to be mildly dangerous (a better term here might be risky instead of outright dangerous), today's reactors are much safer. The tricky part is, as with most debates, both parties are right to some extent, because while new reactors are extremely safe, the existing ones come with some problems of the past. As the popular astrophysicist Neil deGrasse Tyson is fond of saying, "One of the great challenges in this world is knowing enough about a subject to think you are right, but not enough about this subject to know you are wrong." This debate seems to be the epitome of this philosophy.

While this book has thus far made, and will continue to maintain, the argument that nuclear power has historically been and still is exceptionally safer than almost any other form of reliable energy production and shows small risks even when compared to certain renewables (specifically hydropower and solar), the fact that most of the world's nuclear power facilities are relatively outdated for the modern age cannot be overlooked in any discussion of this topic. This does not mean they are inoperable or unsafe by any means, simply that they are not up to par with the most modern designs. As we saw earlier, many of these plants are still operating well after the initial lifetimes designated.

If nuclear power generation is to have a future—and it increasingly seems that it will be needed—then it will almost certainly lay with gen. IV nuclear technologies. As we explore the opportunities and technologies behind gen. IV throughout this chapter, keep in mind that while these features are split into different sub-sections for the purposes of comprehension, many will work cooperatively with each other.

Small Modular Reactors. Why are solar and wind power so cheap to manufacture and install? Despite using much more material and land, renewables are rapidly becoming increasingly cost competitive in many regions around the world. The reason is because they are mass producible. You cannot build an entire traditional power plant on an assembly line and roll it out to wherever you need it. Every power plant is purpose-built to fit its surrounding. Every site is unique (to some extent) and therefore expensive. Solar and wind technologies do not have this problem. They are mass produced just like anything you would buy at the store, making the economy of scale implications massive. Between 1980 and 2012, the cost of solar fell by 97%.[8] Additionally, solar and wind can fit any size of power generation. You can have an energy farm that is capable of producing 10 MW, 1000 MW, or anywhere in between while utilizing the exact same technology in every instance. This is far from true in most other energy generation sectors. While the typical power plant can adjust its output depending on the needed energy, they do not have the range of production that wind and solar can provide.

No energy industry (possibly no industry ever) is more of a polar opposite example of this than the modern nuclear energy industry. Transforming the current industry—plagued by long construction times and budget overruns—into an industry that can utilize this assembly line type production would be the single biggest change the commercial nuclear power industry has ever seen.

Designs that utilize this methodology are called Small Modular Reactors (SMRs), which are far from an unproven technology. Early nuclear technology was originally designed to be SMRs for use in military applications such as naval vessels and even for powering aircraft (thankfully, that idea didn't take off). When you hear the term "Nuclear Submarine," for example, this is not referring to a submarine carrying nuclear warheads—though many do—but is referring to the propulsion system inside of the submarine, typically a small PWR. Nuclear power has massive benefits for military applications so it makes sense that there would be specific designs for military applications, instead of civilian. Submarines in particularly benefit from nuclear propulsion because they need to be quiet, stay submerged for long periods of time (therefore not needing to refuel often), and ideally not create harmful gaseous emissions for the crew or use up their oxygen supply. All of these factors make fossil fuels a terrible choice, but nuclear a great choice. Because of this, the world's first nuclear submarine was commissioned by the

US in 1954,[9] less than a decade after the trinity test—the first explosion of a nuclear weapon. Ironically, it is this technology that was then adapted for civilian uses, where it eventually turned into the massive, overbearing gen. II and then III reactors we see today. SMRs are still widely used today on submarines and aircraft carriers.

Gen. IV nuclear SMRs will focus on evolving this technology for a more modern use in civilian applications. SMRs will take the advantage of solar and winds manufacturability and apply it to nuclear power. No longer will nuclear plants simply be massive, looming projects. Additionally, traditional commercial nuclear power is only applicable for largely populated areas, so it is often impractical to build them in rural areas. With an SMR, you can order a reactor that provides say, only 100 MW of power for a town. If the town grows, you can just get a second one. And then add a third. And so on. College campuses or very large businesses can have a small, onsite reactor that provides them all the power they need, 24/7, 365 days a year. This means that the adoption of commercial SMR technology will expand nuclear powers market potential significantly. This has benefits outside of just availability and economics.

For example, a major storm can take out power for massive regions when it forces a plant to shut down. Centralized systems of energy production also are at major risk to terrorism and cyber-attacks. If all of the electricity going to NYC for even a few days was disabled, the economic, social, and humanitarian consequences would be catastrophic. Chaos would ensue rapidly, and thousands could die. Destroying the Hoover Dam would essentially render Las Vegas inhospitable. The city would be reduced to a fraction of the population and would not thrive as a major city any longer. The decentralization of electricity production through wind, solar, and SMRs brings the national security risks associated with the modern power grid down dramatically. Additionally, in the case where a major storm knocks out a power grid for a few weeks (such as with Hurricane Katrina), a couple SMRs could be quickly brought in to provide electricity while infrastructure is being rebuilt. This will only become more and more important as climate change worsens. Situational responses like this are already being seen with the use of solar panels after a major storm, but they come with their own limitations for this application. A massive quantity of solar panels is needed to equal the power of a single SMR, and they only produce power intermittently. SMRs can be used to rapidly restore power to key infrastructure such as traffic lights and hospitals.

In a 2013 TED talk, Taylor Wilson makes various claims about a specific type of SMR that he is working on at the time. These include the need to only be refueled once every 30 years, the ability to transport over highway on standard cargo trucks, and assembly line type fabrication,[10] though there does not seem to be any noticeable progress made on this claim. While Wilson's idea was indeed radical and seems that it may never come to fruition, there are similar projects that may be getting close to realizing SMR technology. The US DOE has recognized the opportunity for this technology and offers a plethora of funding.[11] Largely due to these incentives, there are numerous SMR designs being worked on,[12] and many of these look promising for commercial starts by the late 2020s.

Molten Salt Reactors. Molten Salt Reactors (MSR, I will try to minimize confusion between MSR and SMR) are an ingenious type of gen. IV technology that is becoming extremely popular in gen. IV reactor designs. In a typical nuclear plant, the heat created by nuclear fission is used to boil water, which then turns a turbine. This is fine and dandy, but when an error occurs this simple process causes a plethora of problems such as steam flashing, hydrogen buildup, and of course the melting of a reactor. On top of this, these safety concerns are what are largely responsible for the increasing costs and construction times behind current nuclear technology.

MSRs solve these problems by inherently changing the cooling system used for nuclear power. Instead of using solid nuclear material, the fissile fuel is dissolved into a liquid salt. This allows it to not only fission more easily, but since the material is dissolved in the liquid salt, there are not concerns of a meltdown (it's already melted!) or hydrogen bubbles forming. Liquid salts also have high boiling points, which allows the reactor to easily maintain its process but makes it very difficult for it to reach a point where the reactor has problems with pressurization or the liquid turning into gas, as can be seen with conventional reactors. Most importantly, this high boiling point allows the reactors to operate at very high temperatures, which means they can more efficiently generate power. The higher the operating temperature, the higher the efficiency of the system. They even operate close to atmospheric pressure, which further minimizes safety concerns and cost of manufacturing.

In the case of a loss of power, a "freeze-plug" at the bottom of the liquid flow that is normally kept cold will lose cooling power and melt away. This then allows for the liquid solvent to drain, by gravity, safely into dump tanks where the salt is rapidly cooled, and the fission process is stopped. This is not possible in traditional reactors because the fuel is solid, although gen. III reactors have implemented similar design features to some extent on some reactors. Because of their design, MSRs are inherently small and therefore modular. MSRs also have large potential for use of fuels other than uranium or plutonium, and the ability to "breed" fuel, as is discussed later.

On top of all of this, MSRs self-regulate their temperature, helping to prevent overheating. Since the fuel is liquid, when it gets hot the liquid expands. When liquids expand, the atoms get farther apart from each other, and this means that the likelihood of a neutron creating a fission reaction actually gets smaller as the temperature increases. Since less fission occurs, the reactor automatically cools back down.

I have mentioned several times throughout this chapter of how advanced technology for use in civilian applications was sidelined for military interests; MSRs are the prime example of this. The MSR was developed in the '60s at Oak Ridge National Laboratories—where most of the early US nuclear innovation occurred—through the Molten Salt Reactor Experiment (MSRE). The MSRE at Oak Ridge operated for almost five years (January 1965 to December 1969), and clocked in more than 13,000 hours at full power.[13] In a 1968 review of the MSRE, Murray Rosenthal, the program manager for the MSRE, makes the following assessment of the programs progress:

> "We felt in 1965 that the molten salt reactor had great promise as a safe and economic breeder and that operation of the MSRE was a matter of *national and world importance*. Now, three years later, we believe molten salt reactors look even more attractive, and our confidence in their prospects has increased. This confidence is based partly on our experience with the MSRE and partly on some very favorable developments that have occurred during the past year."[14] (Italics added.)

He then concludes this report by saying:

> "…we think that it will turn out to be workable and economic, since the concept is flexible and the equipment for the liquid-liquid extraction system appears to be small. The major tasks that face us now, and on which work is in progress, are the engineering of a larger reactor and the development and demonstration of the reprocessing plant. From here we hope to go on to the construction of a breeder reactor experiment that we believe can be a stepping stone to an almost inexhaustible source of low cost energy." [15]

The MSRE program would then be retired so funds could be sent to the pursuit of a different type of reactor that produced more weapons grade plutonium.[16]

Thorium Fuel Cycles. Currently, we use uranium or plutonium (Pu) to power our nuclear reactors—with the large majority being uranium. Aside from the waste created through the nuclear fission process, uranium mining can be hazardous to both the environment and populations near the mine, not to mention how expensive it is. Through the mining process of uranium, vast amounts of radioactive debris and sizable amounts of radon gas are unintentionally removed from the Earth and into the air. These have been shown to greatly affect those who are working and living in and near the mines. Cancer rates from radon and the release of radiation from the uranium have been strongly linked to those who lived downwind of the mine over extended periods of time.[17] Although we need much less uranium for a power plant compared to coal, if we want to truly produce clean energy for the world, mining is also going to have to play an important role in the future.

While uranium is actually a common metal in the Earth (even more common than tin), the specific isotope of uranium that is used for nuclear reactors, called Uranium-235, is very rare. U-235 only makes up roughly 0.7% of naturally occurring uranium (the rest of which is U-238),[18] meaning that it's more than three times rarer than silver in nature. Because of this lack of fissionable uranium, we dig up large amounts of uranium and through a process called "enriching" are able to increase the concentration of U-235 to a level high enough for the fission process in a reactor to take place. As discussed earlier, 3-5% of the uranium in a sample used for fission power is comprised of U-235. For nuclear weapons, this amount needs to be about 90% and is one of the biggest reasons why it is so much harder to create nuclear weapons instead of nuclear power.[19] This 90% enriched uranium is what you may have heard called "weapons grade uranium" in the past, and is exceedingly difficult to produce.

Aside from the horrible mining effects, the long lasting radioactive waste, and the expensive enrichment process, uranium is great! I say that sarcastically, but in all seriousness it creates massive amounts of energy for us to use, the reactors give off no greenhouse gases, and much less of it is needed than other fossil fuels. It turns out, however, that there is a way to get all of the positives of nuclear fission, without the negatives. It comes down to changing to a better fuel. Although uranium and plutonium have helped us greatly, it may be time to find a replacement. The candidate for this is an element called thorium (Th).

Thorium sits just two seats down from uranium on the periodic table and was named after the Norse god Thor. Thorium based fission reactors have the potential to make gen. IV nuclear technology even better than has been discussed thus far in a major way. It can provide us with much more energy, that is much more obtainable, easier to fission, produces far less radioactive waste, and is significantly cheaper than uranium fuel, all without losing sight of the positives of conventional nuclear fission such as the massive energy density. The MSRE discussed earlier was built as a steppingstone for a thorium fuel cycle experiment. The reactor ran on U-235 and a small amount of U-233 towards the end (the U-233 was created from thorium, though it never contained thorium inside of the reactor). The next iteration was going to use thorium, but of course, the program ran out of funding.[20,21] In addition, the Shippingport nuclear station (one of the very first large-scale plants to generate electricity) utilized a thorium fuel cycle for several years. Thorium can be used with several different types of reactors, but the most promising is the use of thorium in MSRs. Not only is MSR technology very promising by itself, but the add on of thorium to the fuel cycle could change how we view nuclear power on a fundamental level.

Thorium is cheap, very cheap. This is mainly because it is such a common element; thorium is about as common as lead, and it's also much easier to mine and less environmentally damaging than uranium. Unlike uranium, there is only one naturally occurring isotope of thorium, Th-232, which is the isotope required for fission.[22] This means that Th-232 is more than 400 times more common in the Earth than the U-235 needed for today's fission.* With the combination of possible assembly line type reactors, easily accessible materials, and cheap fuel, thorium reactors could potentially dominate the power industry in the future.

Although, Th-232 itself is not naturally a fissile material. It is, however, a fertile material, meaning it can become fissile through a process. It turns out that when Th-232 absorbs a neutron, it converts into Uranium-233, which is a highly fissile isotope not found at all in the natural world.[24] Through this process, we are able to produce very efficient reactors. Since essentially all of the fuel material in this would be fissile, the efficiency and power generated from these types of reactors would be much greater than those using plutonium or uranium. Additionally, as the U-233 is burned, it can create a chain reaction, very similarly to how current fission reactors work. The difference here is that this chain reaction can turn more thorium into uranium and cause fission in that uranium. This means that it is comparatively easy to convert all the thorium into fissile U-233 unlike the enriching process of current fission plants. While this will cause the proportion of fuel that is fissile to go above the previous 3-5% enrichment in some designs, it will still not be weapons-grade.

Thorium was ignored because it was not good at producing weapons grade nuclear materials. The reason is because in conventional reactors there is a lot of U-238 sitting around, not helping with the fission process. They are not, however, doing nothing. When U-238 absorbs a neutron, it eventually turns into the fissile, Pu-239, the ideal material for nuclear weapons.[25] It can only be manufactured at large scale in the presence of U-238 inside a nuclear reactor. In a thorium fuel cycle, uranium is not mined up from the earth, meaning that there is

---

* Here's how this calculation was done: Uranium comprises 4 Parts Per Million (ppm) in the earth's crust, and Thorium is 12 PPM.[23] If only 0.7% of uranium is U-235 this means that U-235 makes up .028 PPM. 12/.028 = 428.57 times more common.

no U-238 and therefore no weapons grade plutonium that is created. This was a big downside to the US military during the Cold War, but is a massive upside amidst increasing worries of nuclear proliferation today. You still can get weapons grade material from thorium fuel cycles, just not plutonium. (U-233 is an exceptionally effective material for nuclear weapons.) However, the fuel cycle also produces U-232 along with the U-233. U-232 is notorious for being an obstacle in nuclear weapons development, often being called "bomb poison", and because it is the same element as U-233 they cannot be chemically separated from each other, rendering the mixture useless for nuclear weapons. There are still ways to separate them (which I will not worry about explaining) but know that the presence of U-232 provides an appreciable obstacle to weapons proliferation in a thorium fuel cycle.

Moreover, a number of thorium reactor types, particularly Liquid Fluoride Thorium Reactors (LFTR, a specific type of reactor that is both a Breeder Reactor and an MSR), are capable of taking the nuclear waste that they produce, and running it through the reactor again. Reactors utilizing this have been proposed with other fuel sources (including nuclear fusion),[26] but thorium seems to be the most applicable near-future candidate for this. This "second go" allows the waste from before to further breakdown and produce more energy. Simultaneously, this greatly reduces the amount of radioactive waste and decreases the half-life of the waste. In some cases, nuclear waste from these reactors could even be run through multiple times, and we can use existing nuclear waste from today as fuel—massively reducing the quantity we have.[27] Breeder Reactors are not a new technology but are only used in limited capacity around the world because of their increased costs when compared to the simpler PWRs or LWRs.* Breeders were supposed to be a key technology that would bring the world infinite clean and cheap power from nuclear, but it simply never took off.

Thorium reactors have the complete potential to be the bridges between fossil fuels and more advanced forms of energy like solar and fusion. After being largely forgotten for over 40 years, much more

---

* If interested more about Breeder Reactors, there are a plethora of scientific papers discussing their potential and the even greater impact that adding LFTRs can have to this technology.[28]

attention has been brought back to thorium. India in particular has shown extensive interest into thorium nuclear power, likely since they have some of the world's largest thorium reserves, along with Australia.[29] China also has a thorium reactor program, though the Chinese government is very protective over their nuclear developments so the seriousness of the program or how much progress has been made is difficult to tell. Due to how cheap and common thorium is, it has the potential to provide us energy for an extensive amount of time, and because it is so common, many mining industries simply threw it away when they would come across it—which was common—because it doesn't have any particular uses outside of nuclear reactors. Now that there is greater interest in thorium, many countries have bothered to begin passively collecting thorium when it is mined up with other resources. France, in fact, already has about 10,000 tons of thorium stockpiled. Estimates show that the country may only need around 60 tons a year for all of its electricity, so the French could already have enough fuel stored for well over 100 years of its power generation![30]

Furthermore, the Moon has large amounts of thorium, meaning Moon colonies could rely on this technology in the future as a source of power.[31] (Solar power could provide consistent power some of the time, but there are periods of time where they would be exposed to darkness. Without an effective storage system, solar power will not be very effective on the Moon. There is—contrary to popular belief—no "dark side of the Moon." In fact, neither side of the Moon is any more or less dark than the other.[32] NASA plans to establish a moon base by 2024 using a uranium fueled SMR design due to this limitation of solar power.)

It may sound fantastical, but a ball of thorium small enough to fit in your palm contains enough energy to provide you with all of the energy that you use in an entire lifetime. And as one engineer put it, this includes the energy you need "…not just for electricity, but everything. For heating my house and cooking my food and for building roads and schools and houses and hospitals, and to manufacture all the goods and products that I need throughout my entire life." [33] Simply put, thorium hasn't quite missed its chance yet, and it could provide a highly advanced energy source. In all, thorium energy still requires more research, but the technology is showing to be very promising and could prove to be an essential part in our transition off fossil fuels. The biggest hindrance to thorium at the moment is that there is simply little public backing (likely due to lack of public education on the subject) and there currently exists no major market for the fuel. Thorium is especially enticing because it seamlessly incorporates small modular and molten-salt technologies (as well as others) together.

Gen. IV Conclusions

At the risk of sounding repetitive, gen. IV technology will have massive implications in the future. It is simply a better way to produce electricity than any other way we have developed to date. As with the other technologies in this book, this chapter only covered a small amount of the information behind gen. IV technologies to bring attention to the most prominent ideas. Some other examples of promising gen. IV reactor designs include gas, sodium, or lead-cooled, supercritical-water-cooled, pebble-bed, very-high-temperature, and newer breeder reactors.[34] What is so frustrating is that some of this technology was already developed and nearing maturing levels over fifty years ago. The world could be so much better in ways that are truly incomprehensible if governments (the US in particular) had not canceled funding for these technologies. But hey, maybe then we would all be speaking Russian.*

Gen. IV reactors will likely still have a major role to play many decades after fusion is created. The most likely scenario currently seems to be that fusion will take the place of modern nuclear reactors—massive, high-budget, expensive, long time-table projects—and gen. IV nuclear reactors—along with renewables—will take up the role of replacing fossil fuels and act as a quick and cheap solution to climate change. There are increasing incentives for next generation nuclear technology due to not just climatic factors but also simply in order to keep the nuclear industry and know-how alive. Programs such as the DOE's Advanced Reactor Demonstration Program (ARDP),[35] and increasing incentives by international organizations such as the IEA are encouraging

---

* I highly doubt it.

for the future of this technology.[36] 2020 Presidential runner Andrew Yang had even proposed a $50 billion investment into R&D for thorium and nuclear fusion during his campaign.[37] President Biden's *Plan for a clean energy Revolution and Environmental Justice* specifically mentions "small modular reactors at half the construction cost of today's reactors"[38] and his proposed *American Jobs Plan* (which focuses on clean energy initiatives) repeatedly mentions the need for development of gen. IV reactors and puts aside large amounts of money for this endeavor.[39]

In fall of 2021, the French President Emanuel Macron reinstated that France will continue to rely on nuclear into the future, and will invest heavily into SMR technologies and nuclear for use in the production of hydrogen fuels. France is not alone though. TerraPower has plans to build its gen. IV Natrium reactor (which uses sodium cooling, a process largely not discussed in this book) in upcoming years, and China and India are investing heavily into Thorium Molten Salt Reactors.

Next-gen nuclear technologies (especially SMRs) will have the added advantage over renewables (with the exception of geothermal and concentrated solar power) in that they can be used to provide both heat and electricity. Fossil fuels are used extensively in heavy industry to provide heat for smelting and other processes, so in order to replace them the common trend looks to be aiming towards electrifying every process possible. However, allowing for nuclear, CSP, or geothermal to provide heat instead of redesigning entire industries around electrification would allow for a much more rapid and less disruptive decarbonization of the industrial sector.

For potentially being so cheap and easy to produce, as well as safer and more resistant to proliferation, gen. IV nuclear reactors could shape up to be quite the bargain and may fundamentally change the way that the global energy sector operates. This is likely why many countries have over the past decade renewed their interest in the technology. As mentioned previously, India and China are showing increased interest in this technology, but so are the US, Israel, and others, including a phalanx of private interests—as is also being seen in the field of nuclear fusion.

Chapter 11

Risking Destruction
Nuclear Proliferation Risks

The relationship between nuclear energy production and nuclear weapons development is one commonly misconstrued. Since nuclear energy was essentially an afterthought of nuclear weapons, it's forgivable to assume that it would be easy for a country to develop nuclear weapons once acquiring nuclear power. After all, as was made clear previously, nuclear power plants were originally designed—at least in part—with the intention of producing weapons grade materials. Because of the inherent and substantial differences in nuclear power generation technologies, this chapter will divide them accordingly between current nuclear technologies, generation IV nuclear technologies (which include reduced risk of proliferation as a mainstream and prominent feature), and nuclear fusion, along with other factors that affect nuclear proliferation risks such as socio-economic and regional instability risks.

The addition of nuclear power in more countries around the world obviously raises several concerns about the possible development of Weapons of Mass Destruction (WMDs)*, but we will focus on answering three distinctly important factors throughout this chapter: How easy is it for an entity with access to nuclear power to create a WMD? ; How likely is it that said entity would want to create a WMD if given access to nuclear power? And lastly; How difficult would it be to stop this entity from creating a WMD, if desired?

These will be important to keep in context and as we go along. For example, a terrorist organization may have a high desire to create a

---

* WMD's can refer to a host of various weapons, including biological and chemical weapons, but for the purposes of this chapter, WMDs will be in reference to nuclear, thermonuclear or radiation based—often called "Dirty Bombs"—weapons.

WMD, but essentially no way of doing so, and would be (relatively) easy to stop. On the other hand, a large country would be much more difficult to stop from creating a weapon but may have no incentive to do so in the first place.

Were Nuclear Weapons Inevitable?

During the second World War, several countries other than the US were working on atomic weapons. This included the Soviet Union, Germany, and Japan, while the UK and Canada worked alongside the United States. Though the US was obviously the first to successfully develop this weapon, the situation could have been heavily in Germany's favor if not for the Nazi party's hostility towards civil service roles, in particularly higher education. The most notable action taken against higher education professionals was through the *Law for the Restoration of the Professional Civil Service* that was put in place shortly after Adolf Hitler seized office.[1] This legislation effectively politicized civil services by excluding so-called "enemies" of the Nazi regime from participating in civil services. Because of the need for scientific research to not be politicized—something that is becoming increasingly common again today—and because of other political and sociological factors, these hostilities and legislatures would rapidly cause many professionals to flee from not only Nazi Germany but also Mussolini's fascist Italy.

Among many others, these refugees included Albert Einstein,* Max Born—of whom Robert Oppenheimer, an American physicist who is now commonly referred to as "The Father of the Atomic Bomb," studied under—and Enrico Fermi—who is commonly referred to as "The Architect of the Atomic Bomb" and was responsible for the creation of the first nuclear reactor. Oppenheimer and Fermi would both go on to play instrumental parts in the development of the atomic bomb through the US's Manhattan Project. Two other notable German physicists who worked on the Manhattan Project were Klaus Fuchs and Hans Bethe and there were a plethora of other scientist refugees who

---

* While Einstein did not help directly in the Manhattan Project, he did help initiate the project through a letter to President Franklin Delano Roosevelt.[2]

worked on other military projects for the Allies.

While many of the scientists who worked on the Manhattan Project opposed the use and development of nuclear weapons—some in fact, briefly believed that an atomic bomb would set the atmosphere on fire, destroying all life on Earth, which was eventually given a chance of just 1 in 3 million,[3] * and of whom, one scientist actually thought it had happened for a moment during the Trinity Test [4]—but were ultimately extremely concerned, justifiably, that if Germany developed the super-weapon first then the Nazi regime would use it indiscriminately to lay waste to all opposing parties with little-to no restraint. Therefore, it was better for the super-weapon to be in the hand of the allies. While Germany was defeated before they could develop an atomic weapon, if the leaving academics in Germany (many of whom were world leaders in their fields) had stayed, the war could have had vastly different outcomes. By the end of the war, it was uncovered that Germany was in fact quite far from creating a nuclear weapon, but it is hard to say what may have happened without a scientific mass exodus in the pre-war years.

Many of these scientists would go on to work on developing peaceful uses of nuclear energy. Because of the need for nuclear weapons through the Cold War, nuclear power was a compromise. It was a way to help both the civilian population—through the generation of cheap and stable electricity—and the military—through the production of weapons grade material and an effective way to power vehicles.

Additionally, several other countries eventually developed their own nuclear weapons. The Soviet Union detonated its first atomic weapon in 1949—just four years after the US. The UK reached the same feat in 1952 (though there was heavy collaboration during the Manhattan Project between the UK and US), and France joined the exclusive club in February of 1960.

Conventional Nuclear Power Proliferation. When discussing nuclear proliferation in the view of an expanding modern nuclear power

---

* Personally, 1 in 3 million still seems like questionable odds to bet the entirety of mankind on. By various estimates, in the US, your odds of being killed by a shark or in a homeland terrorist attack are lower than this.

industry, there are two numbers that you need to know: 33 and 9. First, there are 33 countries around the world that have at least one operating nuclear reactor for commercial power.[5] Secondly, there are 9 countries with nuclear weapons: the US, France, China, Russia, the UK, Pakistan, India, Israel, and North Korea. Of the 33 countries operating nuclear power facilities, the top 5—the US, France, China, Russia, and South Korea—produced 70.5% of the global nuclear electricity supplied, and the US alone accounted for nearly 31% by itself in 2020.[6] Of the 28 remaining countries, only 3 also have nuclear weapons (Pakistan, India and the UK). These countries in turn provided just 3.8% of produced electricity.[7] Both North Korea and Israel have access to nuclear weapons but do not have commercial nuclear power plants.

Summed up, of the 33 countries that possess nuclear power, only 7 also have nuclear weapons. Of these, only Pakistan and India had access to utility-scale nuclear power plants prior to obtaining nuclear weapons. It must also be considered that most of the other countries developed their nuclear weapons programs during the Cold War (US, France, China, UK, and the Soviet Union). In total, there are currently two countries that obtained nuclear power first (Pakistan and India), and two countries that have nuclear weapons but still do not have any commercial nuclear power plants (Israel and North Korea).

Since the process of worldwide denuclearization began, countries around the globe have been taking the enriched uranium and plutonium from decommissioned nuclear warheads and putting them into reactors to use as power. The famous instance of this was the *Megatons to Megawatts Program* between the US and Russia which was completed in 2013, and emphasized the decommission of warheads for use in nuclear power.[8] In 1967, the US had possession of 31,225 nuclear warheads, but in 2019 there were only an estimated 6,185 weapons.[9] This process is unquestionably because of the nuclear agreements and treaties signed, but the fact that nuclear weapons can be used as fuel does provide extra economic incentives to decrease the nuclear arsenal. Not to mention an effective and ethical way to deal with the warhead.

This book has repeatedly mentioned that modern nuclear power in part came from the need to develop nuclear weapons material. This is true, but it can also be misleading if not put into context. While these reactors were originally created for this purpose, this is not their purpose today. This distinction makes modern nuclear plants non-ideal for generating weapons grade plutonium. Because these plants are no longer specifically designed for this, creating and extracting weapons grade plutonium can be very difficult with these reactors. Uranium can also be enriched to weapons grade materials, but this process is time consuming, economically intensive, and can be noticed relatively easily in the modern age of massive surveillance by government organizations and intergovernmental teams addressing the issue of nuclear weapons proliferation (such as the IAEA).

The prime example of this is the US and Israel's intervention of Iran's nuclear enrichment program where a computer virus, dubbed "Stuxnet," was planted on Iranian computers and then proceeded to cause physical damage to the centrifuges used for uranium enrichment (though the actual long-term effectiveness is debatable).[10] This intervention potentially sent back Iran's nuclear weapons program by enough to give international organizations time to appropriately react to the threat of Iranian nuclear proliferation—mainly through the use of sanctions and treaties. The US and Israel had the option of direct military intervention, but at the risk of starting another war, this plan of action was not taken.

When a country is granted permission for the development of a nuclear reactor by the UN, as was the case with Iran, the IAEA regularly checks on the program development to make sure that there are no malevolent intentions behind the use of the technology. This makes the case for a secret development much more difficult. In a "breakout" type scenario—where an entity attempts to create a WMD without hiding it—military responses can be swift but are not always the best call. Sometimes it is better to let a country obtain WMDs than to start a world war over stopping them, as was seemingly the decision made with North Korea.

For countries, the combination of geopolitical influences and the limited technology of modern nuclear fission make developing a nuclear weapon from conventional nuclear power difficult, but not impossible. Other constraints to consider are that with an expanding nuclear power distribution around the world, it would become increasingly difficult for the IAEA to effectively track these risks.

What is likely the biggest threat of proliferation from nuclear plants would be from terrorist organizations. This is because terrorist organizations do not have to worry about geopolitical issues, and because they do not need to spend time developing nuclear weapons. They would almost certainly, if given the opportunity, choose rather to pursue creating dirty bombs from radioactive material. Nuclear plants have notoriously rigorous security, a fleet of armed guards, towers, and various security measures, and in any developed country it is highly unlikely that any insidious organization would be able to obtain radioactive material from a power plant. The issue is that this might not always be the case if nuclear power were to become prevalent in regions that are less stable, such as certain Middle Eastern and African countries.

Fourth Generation Nuclear Fission Proliferation. Many worry that gen. IV nuclear power will increase the risk of proliferation due to both the expansion of nuclear power to more countries that this will likely cause and also because of the decentralization of the energy grid that SMRs may encourage.*

As mentioned previously, addressing proliferation risks is one of the key aspects being tackled under gen. IV technology. As defined by the Gen. IV International Forum (GIF), the official goal for proliferation resistance is that "Generation IV nuclear energy systems will increase the assurance that they are a very unattractive and the least desirable route for diversion or theft of weapons-usable materials, and provide increased physical protection against acts of terrorism." [11] There is extensive research going into many gen. IV technologies, but the MSR specifically has several features that make it exceptionally resistant to proliferation risks. In a GIF report, these included the low inventory of fissile material in the reactor and the diversion of said fissile material from the reactor.[12] It was also pointed out, plainly, in this report that "MSFRs operated on a thorium fuel cycle cannot be used to make plutonium usable for nuclear weapons." [13]

There are various designs being worked on, but overall, the results seem very promising. For many liquid fuel reactors, the mixing of

---

* While SMRs will provide the ability to decentralize grids, they do not explicitly have to. Keep in mind that SMRs can simply be grouped together indefinitely to create a centralized generation location if desired.

different isotopes of various materials makes separating the fissile material that is used tremendously difficult. Most of the SMR designs are looking at ways to safely store the reactor under the ground. This would help protect it from being accessed easily. (Since these designs have passive fail-safes—commonly referred to as "walk away safety"— a meltdown is not a concern). Ways for the reactors to signal that they are being manipulated or interfered with in any way are also being explored, which would greatly help prevent proliferation from non-state organizations by alerting the proper authorities.

As far as concerns over a dirty bomb goes, it is too early to tell what the real risks may be. They could be high or low, but all that can be said in a certain manner is that it is an issue being looked into extensively. Overall, though, it seems plausible that decentralized reactors will require their own security team in any case.

Nuclear Fusion Proliferation. The risk for proliferation with nuclear fusion technology is potentially lower than even that of gen. IV technology. Not only are thermonuclear weapons insanely expensive and difficult to create, but the fusion implosion reaction that occurs in the weapon also first requires a separate, fission reaction to kick-start the explosion, as discussed earlier. This means you can't have fusion without fission, as far as weapons are concerned.

Research from Princeton University into this exact topic determined that the risk for fusion proliferation was very low across all of the scenarios that were investigated.[14] This research looked into the possibility of fusion proliferation in the scenarios that the plant is an undeclared facility, the plant was known but covertly producing weapons, and the case that the plant is openly attempting proliferation. It is important to consider all of these options because they each pose specific challenges. For example, a nuclear plant (of any type) openly attempting proliferation would be able to produce fissile material much more quickly than doing so covertly, but with the risk of immediate oppositional action. In all three scenarios, the study found that fusion energy has a "much lower than the equivalent risk from fission systems, if the fusion system is designed to accommodate appropriate safeguards."[15]

Additionally, because nuclear fusion will almost certainly be massive engineering projects, the likelihood that any but the most economically advanced economies would obtain the usage of this technology (assuming no unexpected breakthroughs in the technology) is massively unlikely. The potential that this technology would be constructed inside of regions where there is ongoing and historic regional conflict is exceedingly low due to the economic implications that come with such a massive investment. In these regions, it is almost certain that the source of energy provided will be fossil fuels, primarily coal, where construction times are cheap and quick. Nuclear fusion may be compelling for advanced economies who are worried about public health and safety—and who will also focus more on long-term growth—but economies going through their own industrial revolutions are typically far more concerned with the rapid growth of their economy that is often looked at from the perspective of a much shorter time scale. Due to the extremely low amount of radioactive material generated by fusion, there is little to no worry about the creation of dirty bombs or proliferation of any sort from malevolent non-governmental organizations. Since fusion will be mainly used in highly developed countries, the possibility they would even desire to create weapons is massively low.

Nuclear fusion is the ultimate non-proliferation form of nuclear power. It is not easy to use a fusion facility for the products of proliferation, countries that will have access would have no desire to do so, and because of the massively complex operations and tight precision tolerances that are required for manufacturing and construction, it would be fairly easy to halt the proliferation process. ITER, for example, is reliant on a massively complex manufacturing and logistic chain that encompasses countries from all around the world. Without the specialized manufacturing abilities of various countries, fusion as we know it is not a realistic power source. If the international body would have reasons to fear a countries abuse of the technology, the supply chain could easily be disrupted.

Socioeconomic Deterrence. Iran is again a good country to look at in the discussion of socioeconomic deterrence. The US has long suspected Iran of a covert weapons program, but it seems as though over the years they've never been able to produce an atomic weapon, if indeed they do/did have a weapons program. As discussed earlier, while there were direct intervening actions taken, this has been greatly constrained mainly because of sanctions and intervention by countries (primarily the US). Iran is also different from most countries. If Iran is developing nuclear weapons, it's because they see it as a survival need in order to defend itself from Israel—a nuclear state—much in the same way that Russia developed atomic weapons after WWII because the US had them.

Because of this reasoning alone, it can be deduced that most countries around the world that do not already possess nuclear weapons would not try incredibly hard to acquire them—even if given the opportunity. This is because it is an awfully expensive and complicated process. The Manhattan Project, for example, was one of the most expensive weapons projects in US history. By the end of WWII, the program had cost the US $20 billion.[16] Comparably, the US spent $24 billion on small arms material for the entire war (both numbers are in 1996 dollars).[17] * Relatively, this is not much for such destructive weapons, but there are obviously many other aspects that now go into development today such as missile technology, and handling of radioactive materials (not to mention the enormous costs that can be associated with trying to produce nuclear weapons secretly).

The real cost, however, would today come in the form of economic sanctions. One 2013 study by the Federation of American Scientists (which, interestingly, was founded by scientists and engineers deriving from the Manhattan Project, Oak Ridge National Laboratory, and Los Alamos National Laboratory) estimated Iran's nuclear program costs to be well over $100 billion.[18] When the nuclear deal sanctions were released off of Iran in 2015, the nation's GDP grew by 12.3% over just the next year. However, when President Trump reinstated sanctions against Iran in 2018 and 2019, the GDP of Iran decreased by 4.8% and 9.5% respectively.[19] In 2019-2020, Iran's oil sector shrank by 38.1% and its GDP fell by 6.8% (also in no small part due the Covid-19 pandemic).[20] Though these sanctions are mostly held by the US, the UN has had a large role as well, and it shows that even the (likely justifiable)

---

* $20 billion in 1996 dollars is roughly equivalent to $33.3 billion in 2021.

suspicion of nuclear weapons development can handicap a country's economy. Obviously, these tariffs on top of the Covid-19 pandemic could have significant long-lasting political and socio-economic impacts for Iran.

It's because of this combination of high costs and international interference that most countries don't have nuclear capabilities. Underdeveloped countries simply don't have money to spend on these weapons programs, and more advanced economies that do try this are met with sanctions that can devastate their economy and take away motive to continue research. The economic (and therein social) tolls all this puts on a country just make it an impractical decision to try to develop nuclear weapons illegally in most circumstances. Additionally, just the risk of Iraq having a WMD was enough for the US to launch a full-scale invasion of the country, quickly overwhelming and defeating the Iraqi military (at least, that is the official justification of the war). The speed and ease of which the US was able to remove Saddam Hussein from power may also be part of the reason why Iran is so concerned about its own defense, and a reason why the country may be seeking WMDs.

Sometimes, a nation-state may see the socioeconomic risks of developing nuclear weapons worth the reward. This is almost always because a country feels threatened by another that has nuclear weapons, so they seek WMDs for self-defense. There are several countries where this can be seen. The Soviet Union developed WMDs to defend from the US and NATO. Pakistan obtained them to defend itself from India. Iran seeks nuclear weapons because it feels threatened by Israel—likely for good reason, though Iran certainly is not kind to Israel either. North Korea's intentions are slightly harder to decipher because the country has frequently acted as the aggressor, not defender. Regardless, North Korea seeks a nuclear arsenal so it can at least be on more even footing with its opponents—primarily the US and South Korea.

The world, however bad it may seem sometimes, is not in a constant state of war anywhere close to what could be seen throughout human civilization. While ancient civilizations thrived off conquering, the world as a whole now seeks peace instead of violence. This is likely the single biggest reason many countries do not have nuclear weapons. Not because they can't afford it, but because they simply don't find them necessary. Furthermore, there is no reason that a country that obtains nuclear weapons cannot give them up at some point in the future. Four countries that at one point had nuclear weapons have given them up: Ukraine, South Africa, Kazakhstan, and Belarus.

Summary on Nuclear Proliferation

The fear of a nuclear Armageddon is not something to be taken lightly. If more countries around the world have nuclear power capabilities, then there is obviously a greater risk of nuclear proliferation, no matter how small—as this chapter has pointed out. Despite this obvious truth, there is evidence that supports the mentality that nuclear weapons actually bring peace to the world. Many historians, researchers, and intellects, such as Steven Pinker and Jordan Peterson, argue that today's world is the most peaceful and prosperous point in human history, though there are also many who dispute this claim.[21,22] However, while

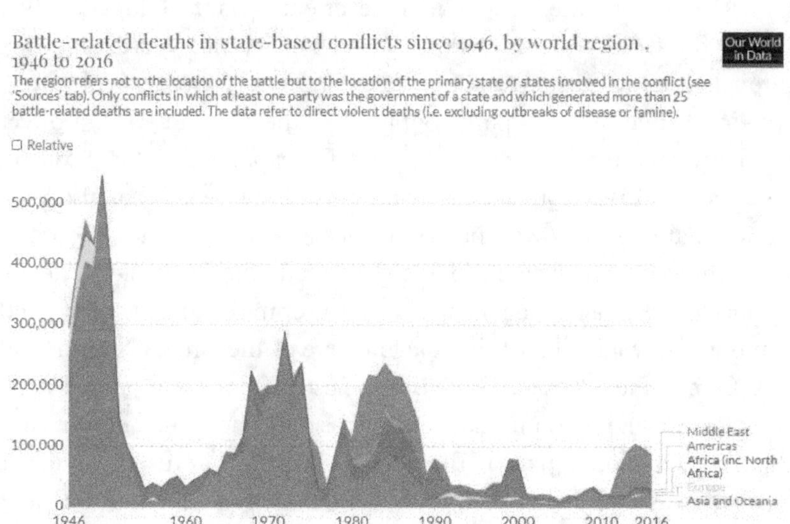

Figure 11: Battle-related deaths in state-based conflicts since 1946, by world region , 1946 to 2016

Source: Roser, Max. "War and Peace." *Our World in Data*, 13 Dec. 2016, ourworldindata.org/war-and-peace#war-and-peace-after-1945.

correlation does not necessarily mean causation, there is a deeply noticeable trend that shows since the creation of nuclear weapons, the frequency and size of wars (particularly between nations who have nuclear capabilities) has dropped significantly (see figure 11), even as global population has more than tripled over the same time.

This could also be explained in part by the forming of alliances and organizations such as NATO and the UN. Additionally, the concept of Mutually Assured Destruction (MAD) explains that opposing countries who each possess nuclear weapons are far less likely to enter conflict, because they could potentially completely destroy each other—leaving no victor. This caused the USSR and US to build their nuclear weapons stockpiles continuously, essentially just for intimidation. Both countries eventually had enough weapons to destroy the other several times over, but never actually used any. As Carl Sagan once explained this, "The nuclear arms race is like two sworn enemies standing waist deep in gasoline, one with three matches, the other with five." It's all just a bluffing game.

Some believe that the concept of MAD is the primary reason the United States and the Soviet Union never entered into a direct conflict during the Cold War. Instead, they preferred to play territorial wars through other nations so as to pursue their interests but also avoid direct conflict. While this did spark conflicts like the Vietnam War, a direct war between the two would have resulted in exponentially more death, destruction and economic costs, nuclear or non-nuclear. It has also been shown that countries who possess nuclear weapons do not choose to use them against rival countries who lack them during conflict. The use of nuclear weapons is one of the worst war crimes that a country could commit and would likely bring the full force of the United Nations and NATO onto the offender. Therefore, the use of these weapons has been completely avoided. A nuclear weapon has never been used in war since the bombings on Japan at the end of WWII. The use of nuclear weapons was available to both the US and the Soviet Union throughout the Cold War, but it was decided not to use them, and these choices gave precedence to the policies and mentality we have today about nuclear weapons.

More countries in the world having nuclear weapons would be scary. And we should continue to closely monitor and restrict any more from obtaining them. However, these factors discussed throughout this chapter show that the likelihood that countries who have access to commercial nuclear power and use this to fuel nuclear weapons development for malicious reasons is not as prominent as one would think off-the-cuff.

Section III

The High Hanging Fruit

Chapter 12

The Problems We Ignore
Introduction to Other Energy Issues

Despite everything gone over in the previous sections of this book, the problems that face the world with regards to energy encompass far more than just those of electricity generation. As many know, a large part of the energy sector is also the transportation industry. This includes planes, boats, trains, and of course, cars. Lots and lots of cars. This is still, however, ignoring large parts of the issue. Industry, for example, is a commonly overlooked problem. Industry primarily refers to construction and manufacturing but can also include other aspects of an economy such as the production of gases and chemicals (which for simplicity we will consider to be the same as manufacturing). The building sector is also a major source of GHG emissions, though it is seldom discussed.

Although the large majority of focus on reducing GHGs gets aimed at electricity generation, transportation is an equally difficult problem. According to the EPA, in 2018 27% of US greenhouse gas emissions came from electricity, but 28% came from the transportation sector and industry produced another 22%.[1] Clearly, transportation is playing a massive part in driving GHG emissions (no pun intended). Additionally, in case you noticed, these three sectors only account for approximately 77% of the US total emissions—affirming just how massive and widespread the issue of unregulated GHG emissions truly is (much of the rest comes from agriculture). Everything we do involves the release of GHGs on not just one, but multiple fundamental levels.

Say you own a company that makes snow globes. Perhaps to offset your carbon footprint you decide to install solar panels on the roof and generate all of your energy with them. Great. Your problem has only just begun. You now have to figure out a way to create the glass for your snow globe a different way since the manufacturing process of glass, metal, and plastics creates CO_2. You will also require a plethora of plastic and other materials for shipping. Most plastics are created from petrochemicals, meaning they are derived from petroleum and natural gas. When burned after use, the chemical breaks down and releases CO_2, or the CO_2 is slowly released as the plastic breaks down over thousands of years. So, to reduce the emissions from these manufacturing process you either need to find an entirely new way of creating the materials or find different materials that will serve the purpose the same or better than before. You could also utilize carbon capture techniques to catch the CO_2 produced through these processes, but carbon capture still has a ways to go before it is an affordable commercially available option (see chapter 16).

Assuming you have figured out how to accomplish these steps, you now have to decarbonize your transportation and distribution. The obvious step here seems to be to electrify your delivery vehicles. So you decide to expand your solar power roof so you can charge your new electric delivery vans with clean energy. Some companies like Amazon have already begun to tackle this. Amazon has partnered with electric car manufacturer Rivian to produce an initial order of 100,000 electric delivery vans, with the whole fleet hopefully on the road by 2030.[2] However, let's not be naïve. Amazon does not drive a delivery van across the country to get you your order. It relies on a massive fleet of ships, planes, and ground vehicles to accomplish this, as well as other delivery services such as UPS. The delivery vans you see every day are just a fraction of the total logistical chain—and undoubtedly the *easiest* to make carbon neutral. Don't be fooled that this is revolutionary. Impressive yes, but there is so much more to do, and 100,000 electric delivery vans is only the beginning of the carbon neutrality challenge. It is also important to consider that the manufacturing of these electric vehicles creates emissions.

Let's assume that you are able to decarbonize your entire delivery system (maybe you just deliver locally, in which case electric road vehicles would suit your needs). The work is still not over. The final large step is going to be to make your buildings as carbon neutral as possible. During the 2020 presidential race, President Trump repeatedly mocked then-candidate Biden's energy policy. One of the main points he targeted was Biden's plan for increasing energy-efficiency in buildings. Despite Trump's concerns, the plan has legitimate reasoning. In 2014, the European Environment Agency noted that 11.5% of total GHG emissions came from the residential and commercial sectors.[3] Also, according to the US Department of Energy's Residential Program Solution Center, updating your home with more efficient equipment and coding can increase the efficiency of a house from 5-30%, resulting in an equal amount of reduction in your monthly energy bill and GHG emissions from home energy loss.[4] Most of this energy loss usually comes from inefficiencies from heating and cooling or poor insulation, all of which make suitable insulation especially crucial and impactful. Even if your house is running on clean power this can still be an issue because refrigerants and coolants produce potent GHGs. Turning in gas or petroleum powered heating and cooling (as many large industrial and commercial buildings have) for electric powered ones will make this process much less painful.

The last piece of the puzzle is going to be construction. As we will see, construction creates unbelievable amounts of GHGs, and it is an exceedingly difficult problem. Since you already had your snow globe factory built, you missed the bus on that one. But keep this in mind when expanding factories or creating new branches and facilities. Low-carbon construction materials may be available in the near-future.

All of this just for a cute snow globe. As we have seen, the release of GHGs goes far beyond electricity production, and many even consider fixing the electric generation aspect of this issue as the "Low Hanging Fruit" and that the real challenge will be fixing these other sectors. Of course, many of the solutions to these issues will be to electrify everything possible. Things that typically are not powered by electricity—such as cars and steel making—will become more and more important to electrify. GHGs are involved in every single aspect of our lives, and fixing this is going to be a major issue. However, these challenges are not stopping large corporations from trying. As mentioned before, Amazon is working on electrifying its fleet, but Apple has also announced its plans to be 100% carbon neutral for its supply chain and products by 2030,[5] while "Big Oil" companies show increasing support for carbon solutions (take what you want from that).[6]

Throughout this section we will dive into what solutions are being worked on, what misconceptions lay behind these energy solutions, and how we can expect the world around us to change as we attack these issues.

Chapter 13

Moving the World
The Transportation Sector

As noted previously, transportation is a massive producer of GHGs. It is intrinsically a part of the industry. There is a popular saying that goes "Humans have never moved forward without leaving something behind." We move our cars forward by burning gasoline to turn our wheels and then leave the gaseous emissions behind us on the road, and our planes and ships do the same in the air and sea. Almost all of the time, what is left behind is a variety of harmful particulate matter and other gases, but what we are going to focus mainly on is the CO_2 that is left behind.

Let's first get a couple terms straightened out before we dive too far in. In the US, we refer to gasoline for our motor vehicles as "gas" even though it is in a liquid and not a gaseous form. Therefore, we will be using the terminology of "gasoline" throughout this book. "Petrol" is the same as gasoline, just the shorthand that is used in various other countries instead of "gas." Similarly, we generally refer to "petroleum" as "oil." "Crude oil" and petroleum are the same, but petroleum will be the main term used. Petroleum can be refined to produce gasoline and other fossil fuels such as kerosene. "Natural gas" is not necessarily petroleum. While petroleum is a liquid, natural gas is in gaseous form and is made primarily of methane (typically 60-90%).[1]

Now that we have our terminology straightened out, we can talk about the transportation issue. Of the 49.4 billion tons of CO_2 equivalent GHG emissions produced globally in 2016, approximately 7.85 billion tons (15.9%) were from the transportation sector. Of this, 11.9% of total global emissions came from just road transportation. Rail, pipeline, air, and ship travel produced 0.4%, 0.3%, 1.9% and 1.7% of global emissions, respectively.[2] Other transportation accounted for another 0.1%. (Keep in mind these numbers are all estimates and are rounded up or down for simplicity. Adding up all of the numbers mentioned here gives you 16.3% of global emissions instead of the real figure of 15.9%.)

Road Transportation

Road transportation by far produces the most emissions out of any of the transportation sectors. This is not because it is the most energy intensive, but just because of the sheer amount of distance traveled every year by road. In 2018, an approximate 3.24 *trillion* miles were traveled on roads in the US.[3] This in turn produced roughly 82% of US transportation emissions.[4] Although road transportation is a large driver of emissions everywhere, citizens in the US notoriously drive more than people in most countries around the world. A study by the US Federal Highway Administration (FHWA) found that per capita, in 2007 US automobiles traveled an estimated 8,761 km annually. This is compared to 6,902 in the UK, 6,541 in France, 4,296 in Canada, and 815 in Mexico. Additionally, one year prior it was estimated that Japanese traveled 1,685 km/capita and Chinese traveled a mere 313 km/capita. The country closest to the US on this data set was Germany, who in 2005, traveled 6,939 km/capita.[5] In total, US automobiles traveled far more than in any other country (see figure 13.1).

There are a few basic explanations for this. First, in case you haven't noticed, us Americans love our "freedom." There are scores of Americans living in large cities that prefer to have personal cars where it would be much more effective to just use public transportation. Many even do utilize public transport, but keep a car around just in case. Asian countries rely heavily on public transportation, unlike citizens in the US, which explains why China and Japan's automobile usage is so low. This vastly increases the average Americans carbon footprint since passenger cars are far less efficient than public transport. Additionally, the US has historically had a very large suburban population, though other countries are increasingly matching this trend, particularly in Asian and African countries where populations are becoming increasingly developed and move from rural areas to cities and suburbs. Additionally, Americans are simply more likely to take a job where they have a long commute than most people in other countries. An hour commute might not be so crazy for most Americans, but it is to Europeans.

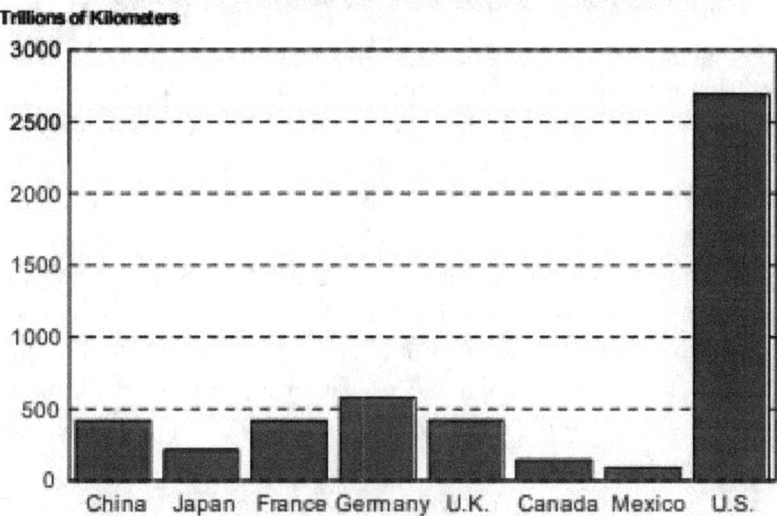

Figure 13.1: Vehicle Kilometers of Travel – Automobile
Source: "VEHICLE TRAVEL BY SELECTED COUNTRY (Metric)." *Federal Highway Administration (FHWA)*, U.S. Department of Transportation, Apr. 2010, www.fhwa.dot.gov/policyinformation/statistics/2008/pdf/in5.pdf.

It is going to be critical in the future that globally there is an increased incentive on reducing our emissions from road transportation. There are several ways this is being looked at, although the elephant in the room seems to be the electrification of vehicles.

Electric Vehicles. To many, the electric car is typically looked at as a new idea. It's seen as one of the great innovations of the early 21st century that is going to save the world. This, however, is far from the truth. The reality is that the idea of an electric car has been around as long as the conventional internal-combustion (IC) engine—what we today commonly think of as a car engine. While the first automobile was created as early as 1769,[6] the IC engine was not invented until the 1860s, with the first practical use cars not coming to markets until the 1880s or even really the roaring twenties. Despite this, Andrew L. Riker had created an electric tricycle (much larger than what we generally think of as a tricycle) by 1896, using it thereafter for a couple years.[7] Additionally, an American by the name of Schuyler Wheeler had filed a patent for the first electric fire engine system as early as 1882.[8] An electric car won the very first American closed-circuit automobile race in 1896, and as late as 1900 there were more electric cars on the road than gasoline cars (and there were more horses than either of these).

Hearing this can be quite surprising. The reality of the electric vehicle is that ever since the IC engine car was developed, inventors, entrepreneurs, and even government organizations were interested in creating an electric-powered vehicle. Throughout the 20th century, companies would pop up and attempt to create a marketable and reliable electric vehicle, but would ultimately fail. One after the other. The brawl between electric and gasoline cars was hotly contested for decades before the IC engine eventually won out. The fact is that, in itself, the electric car is not a new concept, but it was never able to fully succeed until our current century due to limitations in technology—most importantly, battery technology.

When inventors attempted to create effective electric-powered vehicles in the past, they had limited battery options to them. Most of the time they would use chemical batteries such as lead-acid batteries—which is also most likely the type of battery used in your own car (unless you own a fully electric car). With the invention of the lithium-ion battery in 1980 by John Goodenough,[9] batteries—and the electronics field as a whole—were quick to improve in a rapid pace. By the 2010s, lithium-ion technology was in a full-on explosion. As a CNBC article published in late December of 2019 noted, "Over the last decade a surge in lithium-ion battery production has led to an 85% decline in prices, making electric vehicles and energy storage commercially viable for the first time in history." [10] This is because Li-Ion batteries have much higher energy density than other batteries[11] and are relatively easy to manufacture (in part due to the abundance of lithium on the planet).

However, not all of the materials used for these are plentiful. Cobalt, for instance, is not very common on the Earth's surface and much of it is located in the Congo, which makes this a prime resource to become a "conflict material." As reported in 2019 by *The Washington Post*, The Democratic Republic of the Congo (DRC) accounted for 60% of worldwide cobalt production. Additionally, it was estimated that roughly 50% of global cobalt reserves lie with the DRC.[12] As the Congo is an incredibly unsteady country—both politically and economically—this abundance of a much-wanted material makes the country a prime region for forced labor, child-labor, and incredibly unsafe work conditions with little to no workers' rights.[13] Additionally, there are increased worries about after-life processing with these materials, as has been seen with other electronics and PV solar panel waste. And cobalt is just one example of around thirty such rare-earth metals and minerals that are causing huge bottlenecks in the supply chain for much of our technology today, of which many are almost completely controlled by China.

Despite the issues with certain resources like cobalt, from availability and ethical skepticism, electric vehicles are finally seemingly here to stay. Not only that, but they are making the IC engine run for the hills. It is not just that electric cars are a solution to climate change, but in many ways, they are simply a better technology. They are more energy efficient, cleaner, cheaper to maintain, they will make fueling easier (not harder), they are both faster and safer, and are better suited as workhorses.

Electric cars have considerably more torque, meaning that they are better at pulling loads and handle traveling up hills better—a huge advantage for trucking. An electric truck can pull a heavier load much more easily than a conventional IC engine truck. In addition, since electric motors are so much more efficient than IC engines, they will save massive amounts of money and energy by using electricity instead of fossil fuels.[14] This helps to explain why many companies—notably Nikola, Rivian, Tesla, NIO, and many large established automakers—are investing heavily into both the consumer and logistical trucking markets. In early 2020, Rivian held fundraising for its electric truck that was evaluated to be $25 billion,[15] and Tesla has massive financial backing for a company with such a relatively low production-capacity.

Despite this, electric trucks have not quite won the battle yet. The trucking industry seems very prone to the idea of possibly switching to hydrogen fuel-cell technology (which is discussed more later this chapter) instead of going electric. This is because of one main problem: Although batteries are more efficient than fossil fuels, they weigh a lot.

Additionally, because the battery is a solid piece of equipment, the vehicle does not lose weight as it uses energy, like with the conventional IC engine. Simply put, as Electric Vehicles (EVs) use energy stored in their batteries, they carry around more and more dead weight. Not only this, but the battery packs seen in proposed electric trailer trucks take up large amounts of space, meaning there is less room in the truck overall for cargo. These barriers are difficult for trucking, but impossible for airlines and shipping.

Because of Tesla's early breakthroughs in the electric car industry (along with established car-makers early refusal to take the electric car market seriously), the company's market cap at the time of this writing is a staggering $1 trillion, and at one point "Tesla's market cap [had] increased by more than $500 billion … and [was] worth as much as the combined market cap of the nine largest car companies globally."[16] Furthermore, this caused Tesla CEO Elon Musk to become the world's most wealthy individual, surpassing Jeff Bezos for a period of time. No longer is Tesla scared of being put out of business by the established players, but the tables have turned, and established powers are now scared of being run out of business by Tesla and other EV manufacturers. Despite this, in March of 2019, EV market share of car sales was only approximately 1.8%,[17] meaning that there is still a long road to travel before the majority of cars on the road (at least in the US) are electric.

In addition to having a large amount of torque and higher efficiency, EVs have fewer moving parts, making them less likely to break down or need repairs. Also, since there is no engine, EVs are generally safer during an accident (especially from head-on impacts because there is a much larger crumple zone in the front of the car). Since there is no gasoline in the car, the chances of a fire in the event of a crash are greatly reduced. (Although rarer, electrical fires can still occur in an incident with EVs and are typically harder to put out than gasoline fires.)

The market cap of Tesla and other car manufacturers shows that the electric car industry is set to explode over the upcoming years. Further incentivizing this transition are governments from around the world. California has made plans to ban the sale of new gasoline-powered cars by 2035,[18] and countries with plans to ban gasoline vehicles include China, the United Kingdom, France, Germany, Denmark, India, Japan, Spain, South Korea, Taiwan and more.[19] There are still issues with current battery technology—such as degradation of the battery over time—but solid-state batteries are likely to radically improve on this technology in the coming years. Solid-state batteries could also drastically reduce the charging times of electric vehicles, though these concerns are often exaggerated anyways. If it takes you two hours to charge your car, that obviously worries many about wasted time. However, most cars are idle for the vast majority of their life (90-95% of the time by some estimates). Electric charging stations take advantage of this idleness by allowing you to charge the car while you are otherwise occupied. EVs could be charged while at work, eating out, while inside the grocery store, or at home. EVs will not cost time, but will save time by not needing to go to a gas station. With many having ranges nearing 400 miles, there are very few people who the range would hinder. The implementation of solid-state or other emerging technologies could also greatly increase this range.

But why did it take so long for EVs to become popular? And why is it that it was not an established automaker—with countless billions of dollars at their disposal—who finally made the breakthrough? Well, the answer is not clear, but some prefer to get a bit conspiratorial. In the documentary *Who Killed the Electric Car?* the question of why the electric vehicle market disappeared in the late 1990s/early 2000s is investigated. As pointed out in the film, General Motors had a massively successful, small scale, electric vehicle—the EV1—but then decided to give up on the project despite a large demand. Ultimately, the film concludes that Big Oil, Big Automotive, and the government were in a quiet coalition to prevent the expansion of electric vehicles, as it would hurt the oil industry.

It seems all but certain that at this point, the typical battery-pack electric car will be the mainstream replacement for the fossil-fuel-powered road vehicles for today. Nevertheless, there are still operations underway that are exploring alternative, carbon-free fuels. The main reasoning behind this is that there are those that remain skeptical of the potential of EVs. Several factors are at play here, but the main concerns are environmental and/or social impacts of mining for materials in batteries (such as rare-earths as discussed before), a disbelief that the consumer market will shift over towards electric because of concerns such as charging time, and a worry about the implications that an increasingly electrified vehicle market will have on current electrical infrastructure—which may limit the technology to high density urban regions at the start.[20]

It is also worth mentioning that EVs produce no local air pollution, whereas combustion vehicles dump their pollutants directly in the middle of populated areas. This, of course, then goes onto disproportionately affect people of color or those who are in economically challenging positions to a much greater extent than richer, outer-city populations. Even when an EV is powered by a dirty fossil fuel, coal for example, the resulting pollution is emitted relatively far away from populated areas. And even if an EV is powered by the dirtiest coal, its CO_2 emissions become negligibly different to those of current gasoline engines, all whilst creating extensive benefits to the health of communities. To sum it up, in the worst case, EVs save lives.

Hydrogen Fuel-Cells. Several companies are looking into the possible introduction of hydrogen fuel-cell technology into road vehicles, with a few already on roads. Many consider the adoption of fuel-cell technology to be the electric-battery powered vehicles only realistic future competition (aside from existing gasoline vehicles obviously). The main reasoning behind this is because fuel-cells are also electric. Because of this, the technology provides many of the same basic benefits of battery EVs with the additional benefit of using liquid fuel. This means that there is no charging required and we can continue to use gas stations (which would need repurposed to use hydrogen fuel instead of gasoline) in the future instead of relying on long charging times. There have already been various cars made with fuel-cell technology (not all using hydrogen as the fuel),[21] but for various reasons regarding the process and needed infrastructure, it is not clear if it will have a role for road vehicles in the future.

The biggest hurdles for this technology are the massive infrastructure changes required and the inefficiency of the process in getting hydrogen from a source of fabrication to the gas station. Hydrogen technology is further explored with regards to aviation later this chapter.

Advanced Biofuels. There has also been extensive research put into the possible application of advanced biofuels for energy needs, including but not limited to application in road vehicles. Historically, "Big Oil" has been leading the charge on biofuel R&D (as well as hydrogen). As self-reported by ExxonMobil, the company has spent approximately $250 million into biofuel R&D over the past decade.[22] However, biofuels and biomass still produce GHGs when burned. The idea is that they sequester the same amount of carbon when they grow as they give off when they are burned, but there remains a raging debate over whether they are actually carbon-neutral or not. In the very real possibility that they indeed are not carbon-neutral, this makes them not ideal for use as a transitional fuel and is primarily the reason they are not taken as an equally serious solution as hydrogen or EVs. Although there are clear benefits to biofuels, it seems unlikely they will develop to become a real solution inside of road transportation. Like with hydrogen, biofuels, if developed to a mature level, are more likely to be used as an alternative fuel for aviation or shipping instead of road transportation.

Natural Gas. Natural gas has also been explored as an alternative fuel for this sector. As with biofuels, natural gas vehicles (NGVs) still emit GHGs. Despite this, NGVs have gained a foothold. There are already approximately 23 million NGVs on roads across the globe, roughly twice the number of battery-electric vehicles.[23] One large advantage of NGVs over electric and biofuel vehicles is that existing gasoline vehicles can be easily converted to use natural gas (for several thousand dollars). Additionally, although there would need to be additional infrastructure installed, various countries (in particularly the US, Russia, and Europe) already have vast infrastructure dedicated to natural gas for electricity production.[24] This can be tapped into to vastly aid the infrastructure transition towards NGVs. Further advantages include reduced maintenance costs, safety benefits, cheaper fuel prices (especially in the US and Russia), and increased efficiency.[25]

Obviously one of the biggest downsides here is that these vehicles still produce carbon emissions. However, there are a few other problems with NGVs. Because of the way the gas is stored and the requirements of keeping it contained, these vehicles have much shorter ranges. For example, the natural gas Honda Civic has a range of up to 250 miles, while the gasoline run Honda Civic has a range up to 520 miles. The NGV industry is also simply not poised for rapid growth. NGVs are more expensive than gasoline cars while exhibiting limited advantages to the consumer. Cheaper fuel and reduced wear are certainly valuable selling points, but in many cases, this is not enough as the vehicles themselves are several thousand dollars more expensive, a problem that the electric vehicle market is on the edge of cracking. NGVs certainly had a chance, but the rate of adoption of EVs is much higher and it is highly likely they will soon outcompete all other forms of mass road transportation. In 2010 global battery electric car stock was just 20,000 vehicles, by 2019 it was nearly 4.8 million, and it will soon surpass the existing amount of NGVs.[26]

Electric Vehicle Infrastructure Concerns. There is worry about the impact of EVs on infrastructure—especially when it comes to the power grid. An increase in electric charging of cars is going to require a ramp up in electric production and other aspects of the electric grid, as well as updating much of our infrastructure actually used to move this power around, both of which only add even more of an emphasis on the need for a reliable and stable energy supply.

As one PEW report noted, "A US Department of Energy study found that increased electrification across all sectors of the economy could boost national consumption by as much as 38% by 2050, in large part because of electric vehicles." [27] This will have dramatic impacts on the energy supply. Since there will be so much more energy required it will be imperative that this energy comes from a clean and sustainable power output. Nuclear, geothermal, and hydro come especially into mind since many of these electric vehicles will likely be charging at night (because of people charging cars in their home garage after work) when wind and solar are at lower production. Additionally, the source matters because an electric car getting its power from a coal plant will obviously indirectly produce more emissions than an EV getting its power from a cleaner source.

Despite this, several studies have found that regardless of the energy source, battery-electric vehicles still produce fewer lifetime emissions than gasoline vehicles.[28,29] Although, a separate study on the impact of emissions in China found that in China's most heavily coal-powered re-

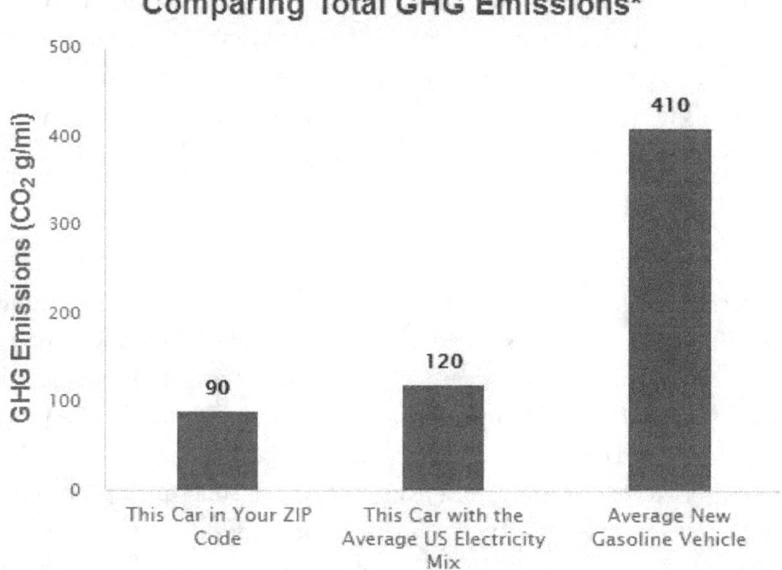

Figure 13.2: Comparing Total GHG Emissions
GHG emissions comparing a 2020 Tesla Model 3 standard range in Washington D.C. to gasoline vehicles. Source: "Greenhouse Gas Emissions from Electric and Plug-In Hybrid Vehicles." *Office of Energy Efficiency and Renewable Energy*, U.S. Department of Energy (DOE), www.fueleconomy.gov/feg/Find.do?action=bt2.

gions, EVs could actually increase emissions by 7.3% compared to traditional IC vehicles.[30] The US Department of Energy has a useful tool to help you figure out what the emissions from an EV will look like for you based on the EV you would be driving and the ZIP code you live in. This considers the efficiency of the vehicle alongside with the sources of power generation available in your area.[31] Figure 13.2 is an example from this tool of the emissions from a 2020 Tesla Model 3-standard range in Washington, D.C (ZIP 20004). The biggest issue with the carbon emissions from EVs is that they produce substantially more CO_2 when manufactured than traditional vehicles, although this is quickly made up for over their lifetime emissions.

There are of course other issues related to infrastructure. Not only will there need to be more net energy produced, but there will need to be charging stations and an improvement of energy grids on a more localized level. Because energy companies have been incentivized to add more charging stations as they produce more vehicles, this problem will likely be gradually resolved as the market widens. Many are hoping this transition to occur within the next decade, but it is more likely this will become a very gradual process, taking around 20-30 years.

Aviation

There aren't going to be any electric-battery airliners for a long time, despite all of the hype surrounding the idea. For all of the progress being made on the electric cars on the road, just about the opposite is being seen when looking at the aviation industry. Lithium-ion batteries just aren't good enough for flying a plane. Because of the "dead weight" a battery-electric plane would need to carry around, planes powered by Li-Ion would simply be too heavy. This causes a massive problem especially for airliners because they need to be extremely light in order to get proper lift, making just getting off the ground from the start exceedingly difficult. They also need to be able to fuel extremely quickly and frequently. Attempting to charge batteries this fast and this often will result in a very quick degradation of the battery. Because of these issues, aviation is likely going to rely heavily on hydrogen or biofuels. Hydrogen has long been in the question as a possible cleaner form of energy. It has the highest energy density of any non-nuclear fuel (by weight), so it makes sense that this would be considered as a future fuel.[32]

For starters, hydrogen fuel would be used as a liquid or gas—allowing it to be pumped quickly, though much of the existing infrastructure for gasoline and oil will not work for hydrogen. Secondly, hydrogen also has only two byproducts, heat and water vapor—making it very appealing for cutting emissions. Hydrogen can be run through a fuel-cell, which essentially takes advantage of certain properties of hydrogen to generate electricity to drive electric motors that would make the vehicle operate (in aviation, this can be used to drive turbines, instead of the jet engines we are now used to seeing) or burned in a typical jet engine like other fuels.

There are concerns and drawbacks to hydrogen, however. Because of hydrogen's properties, it is not commonly found in our atmosphere as a gas like oxygen and nitrogen are. It is everywhere though in the form of water—H2O, two hydrogen molecules and one oxygen molecule. In order to get the hydrogen, we simply split the water molecule apart. There are a variety of ways of doing this, but no matter how you do it, it is an *incredibly* energy intensive process. By some estimates, creating hydrogen fuel is several times as energy intensive as oil production. If fossil fuels are used to create hydrogen, it could drastically increase our dependence on them as well as our carbon emissions, which may be why hydrogen is backed by a plethora of fossil fuel interests.

There are a variety of ways to create hydrogen. Thermal processes use fossil fuels (mainly natural gas) where steam reacts with hydrocarbons to create hydrogen, a process called steam reforming, though this creates large amounts of GHG emissions. Electrolysis can be used to separate hydrogen directly from water, though this requires massive amounts of electricity input. An electrolysis plant run by renewable or nuclear power could produce so-called "green hydrogen" where there are no GHG emissions. Steam reforming is the current way hydrogen is produced (95% of hydrogen is currently created through natural-gas steam reforming), but green-hydrogen through electrolysis is where the industry needs to go in order to combat climate change.[33] Considering that if hydrogen production as of now was its own country it would have the emissions equivalent of Germany, the importance of green hydrogen compensating for expected growth cannot be overstated.

There have long been claims that hydrogen fuel is far more dangerous than conventional gasoline or jet fuels. This misconception is not entirely correct, but it is understandable why one might think this initially. Although the chance is slim, hydrogen can be explosive. If hydrogen gas is ignited, it will rapidly blow up, where as gasoline tends to just ignite into flames instead of being explosive (all in relative terms). Despite this, in reality it is very unlikely to burn or explode because of its lightweight. In the event of an accident, the hydrogen will instantly begin dispersing in the air, and after just a few seconds there will not be any left to continue the reaction. In contrast, traditional fuels stick around and will continue to engulf the vehicle in flames. The disaster of The Hindenburg is often looked towards as an example of why hydrogen is unsafe, yet it was for various reasons, not just the hydrogen onboard, that caused the blimp to explode.[34] There are also various safety precautions that can and are being taken with the on-board storage of hydrogen to mitigate any risks. The reason why hydrogen explosions in a nuclear meltdown are much more of a concern than with vehicles is because there is nowhere for the hydrogen to go. This allows for it to build up in a reactor, and then eventually blow up. In a vehicle crash, there would be many opportunities for the hydrogen to escape. In the case of aircraft, you probably have bigger concerns than a hydrogen explosion if the plane crashes, and it would likely escape anyways.

Likely the biggest threat to hydrogen aviation remains the sheer lack of existing infrastructure and the time it will take to transition to a hydrogen-based fleet. There have been numerous arguments that we shouldn't even worry about the aviation fleet due to its relatively small-seeming impacts and use things such as carbon capture to offset the emissions, but this will be very difficult and could prove to be short-sighted. Aviation is one of the fastest growing sectors. As stated in a 2020 IEA report on aviation, "Since 2000, commercial passenger flight activity has grown by about 2.5-fold (5% per year), while CO_2 emissions rose by 50% (2% per year), thanks to operational and technical efficiency measures adopted by commercial airlines, including new aircraft purchases." [35] Additionally, in 2018, approximately 0.9 of the ~8.1 Gt of GHG emissions that was emitted from transportation came from aviation, which amounts to around 11% of the sectors emissions.[36]

Again, this might not sound like much, but consider that if in 2018 global aviation were its own country, it would have produced the sixth most emissions—just behind Japan. This is more than the emissions of Canada and Italy combined, slightly less than triple of France's emissions, and equal to 90% of total emissions from the entire continent of South America. The only countries that produced more emissions than the aviation industry in 2018 were (in order from first to fifth) China, the US, India, Russia, and Japan.[37] The point is that to simply ignore the issue of transferring the aviation industry to a clean source such as hydrogen could prove to be a massive mistake. Although hydrogen is still very new, expensive, and under a lot of fire, it is also likely the best solution for solving the aviation industry.

If hydrogen is able to take over the roles of larger aircraft, then battery-electric planes (BEPs) can cover the smaller, personal transport area. This will mainly focus on either few or no passenger flights or flights going short distances. For example, the Norwegian Civil Aviation Authority has plans to adopt a fleet of short-haul BEPs, each potentially capable of carrying between 9 to 19 passengers. This in theory should work fine for Norway because the country has "16 airports in radius for short-haul flights of less than 350 kilometers [217.5 miles] on the top of Norway." [38] Norway plans to make 100% of its short-haul civil-domestic aviation electric-powered by 2040.

Outside of Norway and other small nations, the future for BEPs seems bleak. Especially when looking at the US, Americans tend to fly long distances consistently and in large numbers, making this exceedingly difficult to perform. There are many companies working on BEPs, but the benefits from this sector are limited. In 2018, general aviation in the US accounted for just 18.7% of GHG emissions from the aviation sector, while commercial aviation—which utilizes larger aircraft over longer distances—accounted for 74.5%.[39]

Shipping

Between 1990 and 2018, total US GHG emissions coming from the transportation sector increased by 23.4%. However, over this time period, GHG emissions coming from ships and boats decreased by 13.3%.[40] Why is this and how is this possible? First, like aviation, shipping is constantly striving for greater efficiency to drive profits. Operating and fuel costs play a huge burden on international shipping, so it is only natural that these vessels would want to become more and more efficient, thus reducing their emissions, even if that is not the main objective. According to the Third International Maritime Organization (IMO) GHG study published in 2014, international shipping accounts for approximately only 2.1% of global GHG emissions.[41] The same study also noted that "For the period 2007–2012, on average, shipping accounted for approximately 3.1% of annual global CO_2 and approximately 2.8% of annual GHGs on a [CO_2 Equivalent] basis."

Although the share of global GHG emissions from shipping is decreasing, total emissions are still increasing, despite the significant reduction in emissions in US shipping. "Maritime CO_2 emissions are projected to increase significantly in the coming decades. Depending on future economic and energy developments, this study's [business as usual] scenarios project an increase by 50% to 250% in the period to 2050. Further action on efficiency and emissions can mitigate the emissions growth, although all scenarios but one project emissions in 2050 to be higher than in 2012." [42] Because of this, it is of great interest to nations globally to seek out ways in which they can reduce emissions coming from shipping fleets now instead of later.

The European Commission laid out the EU's strategy for reducing GHG emissions from the shipping industry in 2013.[43] Interestingly enough, the report made its case based off of economic benefits, not primarily stemming from the need to reduce overall emissions because of global warming as one might expect. In a short section of the report, the commission made its case by stating the following:

> "In the shipping sector, CO_2 emissions relate to fuel consumed. Reducing CO_2 emissions means reducing fuel consumption which in turn results in savings in fuel costs. As long as the efficiency investments required can be covered by the resulting fuel savings, the sector can earn money while addressing climate change. Such savings are highly relevant in today's context. Fuel prices have been erratic in the last years. They doubled between 2002 and 2005, then tripled between 2005 and 2007, and fallen back to the 2005-level in 2008 just to double again between 2008 and 2010. Prices of heavy fuel oil are now around $650/t, i.e. 8 times more than the average 1990 prices, and they are still expected to increase. Improvements in fuel efficiency have been observed in many segments of the shipping sector only since 2009 when the global economic crisis significantly reduced the profit margin of the sector. ...The impact assessment carried out in the context of this Communication identified progressively increasing savings potential in the cost of fuel that add cumulatively up to €56 billion between 2015 and 2030."

Heavy fuel is a particularly dirty type of petroleum. The continued use of this fuel is the primary reason behind the need to increase efficiency for the international shipping community. The reduction of fuel usage and increase in efficiency has been the main objective for reducing maritime GHG emissions for a long time and will likely continue to be the trend. There have been proposals for alternative fuels such as biofuels and hydrogen to be used with these, but similarly to the situation with aviation, there is still a long way to go before we would begin to see the implementation of these fuels, though hydrogen infrastructure could be more easily implemented for shipping because of its location on the water, where hydrogen could be produced at the docks. There is not much drive for the use of battery technology in shipping (especially in international shipping) due to the massive size of cargo ships, the relatively long time spent out at sea where it cannot be charged, * and the extra weight batteries would add to the design. Of course, maritime shipping fleets could easily be powered with nuclear energy (as militaries have been doing with aircraft carriers and submarines for decades), but this brings forth a plethora of proliferation concerns. For shipping, the name of the game is efficiency.

Rail

While rail represents about 9% of global passenger movement and 7% of freight movement, it only represents 3% of global transportation energy use and is only responsible for about 0.3% of direct global CO_2 emissions.[44] Because of rail transport's impressive efficiency, it is one of the cleanest ways to move around product and people. As an IEA 2020 report noted, "About 7% of global freight transport activity…uses rail. Transporting cargo by rail could be the least energy- and CO_2-intensive way to move freight of any land-based transport mode, but as with passenger rail, its economic and environmental benefits depend upon the long-term certainty of high throughput volumes on certain routes."[45] Seeing as about 29.4% of transport emissions come from just road freight transport, an increased incentive to transport freight by rail could result in massive emissions reductions.[46]

Rail may become increasingly important in the future in relation

---

* Solar power is not applicable because more space than is on the boat would be required. Potentially, turbines underwater could spin as the boat moves, charging up a battery, but this would slow down the boat and make it less hydrodynamic. There have also been proposals for fitting wind turbines onto cargo boats, but the effectiveness is yet to be proven.

to the aspect of civilian travel as well. The onset of high-speed rail (HSR, also commonly referred to as bullet-trains) and maglev trains (a specific promising type of high-speed rail that uses no fuel, only electricity and magnets) could help this in the future. There are many instances where the onset of an HSR can increase not just the use of public transportation, but also actually eliminate the need to fly in many instances. Some of the fastest maglev trains can reach operational speeds of up to 250 mph, but it is more common for these trains to go anywhere between 140-180 mph. Additionally, the US High Speed Rail Association (USHSR) does have several working plans to create HSR lines in the US that operate at speeds of up to 220 mph.

Unfortunately, these systems are almost non-existent in the US for several reasons. First, in the US there are extensive issues concerning private property. HSR pathways need to go in straight lines because they are moving so fast. A turning radius on an HSR can commonly be several miles. Because much of the land in the US is owned privately, this makes construction tricky, lengthy, and incredibly expensive. If one person doesn't want an HSR train going through their property, then large portions of the train's path may need to be redesigned. This has been especially noticeable in California's attempts at creating HSR systems over the past decade.[47,48] While the NIMBY (Not In My Back Yard) mentality has served Americans good for many decades, it also gets in the way of projects like this that can help everyone.

Other countries outside of the US have seen tremendous use of HSR systems. China has been particularly prominent in this transition, while other developed nations such as the EU have also taken great interest in the opportunities presented by this.

> "Passenger rail networks are concentrated in a handful of regions – China, the European Union, India, Japan and Russia – that together register 90% of global passenger rail activity. …Almost two out of three high-speed rail lines are in China: starting from virtually none only a decade ago, the country now has over 24,000 km. In 2019 alone, China National Railways opened two more high-speed rail corridors totalling 750 line km, and added more than 3,000 km of new lines. Owing largely to China's commitment, global high-

speed rail infrastructure expanded 3.4 times between 2008 and 2018. The rate of annual average growth (13%) is historically unprecedented" [49]

Although it is quite unfortunate that countries in North and South America are falling behind in this category, there is a silver lining. As China continues to emit an increasingly massive amount of GHGs, the fact that it is taking these very serious strides that will not only help its economy but also reduce its carbon footprint are reassuring. There are also simple practical and geological explanations for this. HSR systems are often limited by terrain, and in vast areas of the US (especially the east and west coasts because of the Rocky and Appalachian Mountain ranges) this makes building HSR a herculean effort.

Lastly, there is also some slight momentum behind hyperloop technology. Essentially, a hyperloop would take an HSR system and put it into a vacuum tube. This would potentially allow these systems to go much faster since there would be no air friction acting upon the train, which is the biggest single limiting factor to an HSR (especially maglev) systems speed. Popularized primarily by Elon Musk, this idea has gotten increasing attention over the past several years. While this will almost certainly not be a common form of transportation for many decades due to the immense engineering challenges, it could one day be the best mode of public transport, replacing current underground subway systems. There are concepts for this that could result in pods traveling at speeds of over 700 mph.[50] For comparison, a Boeing 747 can only travel closer to 600 mph.

No matter how we do it—whether it's with conventional light and heavy rail, HSR, something advanced like hyperloop, or a mixture of all of them—it is clear that the role that rail transport plays in our lives is going to need to drastically increase. Not only will this be one of the most profitable options due to the high efficiency of rail but will also prove to be a good way to stimulate economies and increase access to healthcare and education in places across the world.

Summary on Transportation

All things considered, the transportation industry is going to need to undergo a dramatic shift over the next several decades, though this is not a bad thing. Although there will be hardships as countries and economies shift to accommodate this, the technologies proposed still lie well within the natural progression of technology—just expedited to deal with the threat of global warming. EVs are better in almost every metric, hydrogen will make airlines and shipping more efficient and cheaper as long as the necessary infrastructure can be created, fuel efficiency standards and alternative fuels (possibly nuclear, hydrogen or biofuels) will need developed and implemented, and rail will not only incentivize more public transport, but will also make traveling quicker and cheaper. The biggest challenge is the development of hydrogen. To comply with the SDS, GHG emissions reductions implemented by hydrogen and electrification of transportation will need to result in 5.06 Gt of CO_2 per year more than the current projections account for, by 2070.[51] (Roughly equal to 93.5% of total US CO_2 emissions from 2018.[52])

Chapter 14

Making The World
The Industrial Sector

In 2018, industrial processes were responsible for creating approximately 22% of the United States' GHG emissions.[1] This equates to roughly 1.47 Gt of CO_2 equivalent GHGs—slightly less than Japan and France's combined total CO_2 emissions.[2] The same year, 24% of global emissions were the result of industry.[3] Despite decreasing by 0.6% that same year, "To align with the SDS, industry emissions must fall by 1.2% annually to 7.4 $GtCO_2$ by 2030 – despite expected industrial production growth." It is important to note that this figure assumes no type of carbon capture and storage innovations, so this might be slightly skewed, but it is hard to say since carbon capture technologies are still underdeveloped (see chapter 16).

Of the 8.5 Gt of CO_2 emissions generated in 2018, 2.3 and 2.1 came from cement and steel/iron production, respectively. Another 1.2 Gt came from chemicals and petrochemicals, while aluminum and pulp/paper each produced 0.2 Gt, and other industrial processes produced the remaining 2.5 Gt.[4] It will be important to note throughout this chapter that many of these industries will naturally become cleaner due to economic concerns as the price of oil and fossil fuels continues to increase, as seen with the shipping industry in the last chapter. For example, petrochemical companies—which are responsible for not only the production of oils and fuels but also for plastics—will naturally innovate as the price of crude oil continues to be driven higher (both by low-carbon competition and future policies such as carbon taxing). For this reason and others, this chapter will focus primarily on steel/iron production and cement use; not only because they are the primary drivers of emissions in this sector, but because these are likely two of the hardest subsectors to decarbonize.

Despite the previous economic drivers mentioned, decarbonizing the industrial sector is not going to be easy. Most of these industrial processes have the production of GHGs (primarily CO_2 generated from burning coal) inherently mixed into the foundational chemistry that is used to create the materials. Additionally, industry is growing at a rapid pace. "The industry sector accounted for 37% … of total global final energy use in 2018 (including energy use for blast furnaces and coke ovens and feedstocks). This represents a 0.9% annual increase in energy consumption since 2010, with 0.8% growth in 2018, following stronger growth of 1.6% the previous year." [5] Like with transportation, much of the industry is going to depend on the electrification of processes, but will also likely require increased usage of carbon capture systems, hydrogen fuel, excess process heat captured from energy production and an inherent change in the way that materials are created.

Cement/Concrete

The most used material on the planet is, probably unsurprisingly, water. The second most used material is concrete. (Cement is the primary ingredient used to create concrete; they are technically different, so this figure would include cement in it. Note also that cement and concrete are sometimes discussed as one in the same.) The sheer amount of cement and concrete produced each year is almost unfathomable. In 2019, there were approximately 4.1 Gt of cement produced worldwide, and China alone accounts for around 55% of global production each year.[6]

Because of chemical processes, about half of the emissions from cement cannot be negated by fuel or efficiency changes as seen with transportation, due to the inherent chemistry attached to the creation of the material.[7,8] This is because limestone (CaCO3) is a primary ingredient, which releases CO_2 when processed for cement. On the bright side, this at least means that the other ~50% of emissions can be solved through efficiency and fuel changes to the process.

The main target in the sights of reducing GHG emissions from the cement industry is clinker. Clinker is essentially the main binding material used to form cement and it's responsible for many of the inefficiencies in the manufacturing process and also the chemistry of creating the mixture. The amount of clinker used in the cement process "…is directly proportional to the CO_2 emissions generated in cement manufacturing" and "From 2014 to 2018, the clinker-to-cement ratio increased at an average of 1.6% per year, reaching an estimated 0.70 in 2018." [9] In order to comply with the SDS goals, the clinker-to-cement ratio is supposed to fall by 0.3% per year, landing at a global average of 0.66 by 2030. Similarly, under the SDS goals, 1 ton of cement would produce 0.48 tons of CO_2 (by 2030), but in 2018 this figure was 0.54 t CO_2/t cement. Again, considering that in 2019 there were 4.1 Gt of cement produced, this seemingly small reduction in CO_2 intensity would have prevented over 240 million tons of CO_2 emissions. This is by no means implausible. China at one point had a clinker-to-cement ration of 0.57, but this has since increased to 0.60. [10]

There are other factors at play here as well. When the cement mixture is being created, it is being blasted in a kiln. The high temperatures produced here often require large amounts of fuel (almost always coal or natural gas) to be burned to create the heat needed, but this is a process that can be electrified. Combining this with the increasing realism of autonomous construction vehicles, efficiency has a high ceiling in construction.

Carbon Negative Concrete. One possible solution for concrete emissions that would be a major game changer would be turning concrete into a carbon *negative* solution to climate change. [11] The basic principle behind this idea is to change the chemistry and mixture of the cement so that instead of using water to cure concrete, it uses CO_2. Some projections of this even go as far to say that not only would a CO_2 cured concrete be better for the environment, but it would also be more durable and easier to use. While this approach is being researched further, there is increasing drive to incorporate carbon capture systems into the industry, which could turn concrete production carbon neutral. The IEA further makes note that "CCUS [Carbon Capture Utilization and Storage] will be crucial to reduce cement sector CO_2 emissions. ... To be on track to achieve SDS decarbonisation, capture technologies in cement production should be commercialised by 2030." [12] There are also many working to create completely new solutions of concrete that does not rely on cement at all and rather uses materials such as hemp or bamboo. Materialistic innovations like these could prove to not only make a major dent in our global GHG emissions but is also a multibillion-dollar market waiting to be exploited.

The World is Running Out of Sand (No, Seriously).[13] The world is experiencing a major sand shortage. As mentioned earlier, the only resource used more than concrete is water. As concrete is comprised 70% of sand, sand mining has become a multibillion-dollar industry.[14] However, let us make something clear, not all sand is created equal. It is not necessarily that the world is running out of sand, but that we are running out of the type of sand that is required for producing concrete, glass, electronics, is needed for fracking, and is used in essentially every single industry. It's surprising how much of your life relies on this simple resource.

> "Sand and gravel are mined world-wide and account for the largest volume of solid material extracted globally. Formed by erosive processes over thousands of years, they are now being extracted at a rate far greater than their renewal. …Globally, between 47 and 59 billion tonnes of material is mined every year, of which sand and gravel…account for both the largest share (from 68% to 85%) and the fastest extraction increase." [15]

The type of sand that is required for concrete is generally the type of sand found underwater, on beaches or below the Earth's surface. Therefore, the type of sand that we find in deserts is mostly unacceptable for our uses (not all sand is created equally). As far as concrete goes, China is easily the single biggest culprit here. Between 2011 and 2013, China used more concrete than the US did during the entire 20th century.[16] However, China is not the only culprit in our sand shortage. Dubai has used gargantuan amounts of sand in expanding its costal real estate through its construction of The Palm Jumeirah, The Palm Jebel Ali, and its "The World" artificial island projects.[17] Singapore also uses ludicrous amounts of sand for expansion of its coast. The Island nation is the world's largest importer of sand "by-far," consuming 5.4 metric tons of sand per inhabitant.[18]

It is a difficult problem to solve because we are so reliant on the resource. There are ongoing talks to regulate sand across the world, but the fact remains that we are depleting aggregate sand much faster than it is being replenished by nature. A priority solution will likely need to be increasing funds into developing machinery and processes that will turn regular rock into sand quickly and affordably. There are also massive environmental and socioeconomic impacts associated with sand mining. This includes biodiversity concerns, coastal erosion, sea level rise, and even a violent sand black market.[19,20] In truth, this problem is so deeply entangled into society that there is little hope for sweeping reforms. In the case that little international regulation is taken up, the global sand shortage could cause future conflicts and may bring the "endless-growth" capitalist society that the world operates on to a highly destructive screeching halt later this century. In the lack of an effective solution to concrete and other building materials that do not rely on sand, infrastructure could suffer catastrophically. It is interesting because unlike with the elements we mine for (such as iron, uranium, and rare earths) sand actually is a renewable resource. We are just using it much faster than it can replenish.

Iron/Steel Production

Iron and steel production is one of the most energy intensive industries. (Energy intensity is how much energy is required to produce an amount of a material.) In 2018, iron and steel production had an energy intensity of 18.6 Gigajoules/Ton (GJ/t)—the lowest intensity since before 2000—while cement production had an energy intensity of just 3.4 GJ/t, meaning it required nearly 5.5 times the energy input to create a ton of iron/steel than cement on average, creating a lot of CO_2 in the meantime.[21,22] The emissions problems associated with creating iron and steel are not necessarily directly tied to the chemical makeup of iron, but they are prevalent in the process that is used for turning iron into steel.

While coal was a necessary part of the chemical transformation of iron ore into steel in the past, today this coal is used primarily for heating and is not inherently needed. Using clean electricity or hydrogen are now being discussed in order to transition to a cleaner methodology.[23]

Coal use is the primary reason why emissions are so high from steel production. In 2018, coal accounted for ~74% of iron and steel energy demand.[24] Therefore, much of the problem with reducing emissions from steel production will be to replace coal with hydrogen to create the steel we need. (Thankfully, unlike cement and concrete, steel is already one of the most recycled materials on the planet and this can help us to reduce our reliance on its production in the future, though even the recycling process creates some GHG emissions.) Despite this, the emissions from steel production continue to grow at high levels.[25] "Global crude steel production increased by 5% in 2018 to reach 1,817 Mt, following 6% growth in 2017. Initial estimates suggest 3% growth in 2019." [26]

Hydrogen Manufactured Steel. The IEA makes it very clear in its 2020 report on iron and steel that sustainably produced hydrogen—sometimes coupled with CCUS—is going to be a major factor in determining the emissions reductions targeted for iron and steel in the SDS. "Innovation will be critical to reduce primary steel production emissions. …Continued efforts [on CCUS projects] will be integral to bring these technologies to full commercialization in the coming decade." [27] Steel produced with hydrogen has the ability to be a major carbon solution in the industry because it produces a similar effect to iron as carbon monoxide—the main driver of turning iron into steel—but instead of creating CO_2 as a byproduct, water vapor is created. In the absence of hydrogen to be used for steel production, various aspects of the process can also be electrified.

Other Industries

Chemicals/Plastics. Chemicals are responsible for significant GHG emissions around the world. A few major examples of high-emitting chemicals are plastics, methanol, and ammonia. The chemical sector is "the largest industrial consumer of both oil and gas" and consumes an impressive 13 million barrels of oil each day.[28] There are limited things that can be done to reduce emissions coming from this subsector, but the main thing will be to stick to the three R's. Reduce, reuse, and recycle. In the case of single use plastics, aluminum, and paper are able to replace a non-insignificant amount of demand.

Aluminum. Aluminum is one of the most energy intensive industries. Despite this, the vast part of the aluminum process already uses techniques that do not inherently create GHGs. The two main processes—

the Bayer and Hall-Héroult processes—can both be used as carbon-neutral solutions if the electricity required is powered by appropriate sources. (While the Hall-Héroult process does create CO_2 as is, this process can be adjusted to remove the CO_2.) Because of its wide-spread reusability, the replacement of certain single-use plastics for single-use aluminum cans (such as in soda cans) can have a great positive impact. Because aluminum mining is relatively benign when compared to things such as heavy metals and uranium, the environmental impacts could be largely net positive if it is properly recycled and reduces our dependence on plastics. Energy efficiency and providing power from renewable sources will greatly reduce both energy intensity (thereby reducing costs) and CO_2 emissions. Aluminum is probably the easiest major industrial activity to decarbonize.

Pulp & Paper. Emissions coming from the production of pulp and paper has been somewhat mitigated by replanting policies in the past couple decades. Despite this, the replanted trees take years to absorb the amount of CO_2 produced at the time a tree is cut down. Paper can also be used to reduce single-use plastics in some cases (such as for use in plastic bags) and a more effective recycling process can help mitigate consumption.* Pulp and paper has limited ways to reduce GHGs (though some aspects can be electrified) so investment into CCUS may be needed to offset emissions.

Summary on Industry

The industrial sector may very well be the single most difficult sector to decarbonize. Not only do many industries have a long established method of business and manufacturing that will have to be changed, but every industry is widely different. While there are many similar factors to deal with in electric generation and transportation, each individual industry is different. Electrification will be possible in many instances, and hydrogen will play a big part in others. But equally as important, both because of the scale and complexity of industrial pro-

---

* However, the usability of recycled paper will hinder this. After just a few times of being recycled (3-5 cycles) paper tends to become unusable anymore due to chemical treatments and degradation.

cesses, CCUS will be critical in this sector. CCUS is not on track for development and implementation in the industrial sector,[29] yet it is likely the single most important factor for reducing our industrial emissions. Without the addition of CCUS, it will be much more difficult to decarbonize all of these industries. (More on CCUS in chapter 16.)

Chapter 15

Living in the World
The Building Sector

The housing sector is another significant contributor to emissions. When including electricity demand and heating, buildings account for 18.4% of GHG emissions—more than the entire transportation sector.[1] Despite this, the carbon emissions from buildings never seem to get much attention. There are a few things that can explain this. First, decarbonizing the electric production sector will simultaneously help to drastically decarbonize buildings, putting the onus on that sector. While many buildings use fossil fuels for providing heat (such as diesel generators and natural gas heating), these heating elements could be electrified and attached to the power grid so as to further decarbonize them. However, this will also put increased stress on electricity grids—especially renewable grids during the nighttime. Because of this, building efficiencies will also need to increase. This means better insulation, more rigorous building codes, and more energy efficient equipment.

Secondly, people don't like it when something directly affects them. It is easy to post on Facebook saying how much you love solar panels. It is far more difficult to rip out all of your windows for better ones, insulate (or possibly re-insulate) your attic, or replace your refrigerator with one that is more energy efficient. Those aspects which impact daily life are incredibly important to the low-carbon energy transition, but nobody wants to go through the trouble. For many, even something such as buying an electric car will be perceived as easier to implement than better building efficiency because of how invasive it feels. CO_2 emissions from buildings reached an all-time high just in 2019,[2] and direct energy-related emissions from buildings in 2019 were 3.1 Gt.[3]

It is for this reasoning alone that buildings are high-hanging fruit on the tree of decarbonization. The technology exists to decarbonize the sector and make it more energy efficient, but most of us don't want it to affect our daily lives. In fact, in many circumstances it makes more economic sense to improve a building's efficiency than to not improve it. While there are upfront costs, these improvements almost always more than pay for themselves quickly.

While there are not many misconceptions about the building sector, a few specific aspects are worth discussing so that you can be equipped enough to have real discussions about this and understand both the importance and difficulty in renovating countless millions of buildings. Because the electric generation sector has been extensively covered earlier this book, the biggest focus of this chapter will be in efficiency improvements.

Appliances and Equipment. As populations grow and emerging economies develop, the market for appliances and equipment in buildings will only continue to grow alongside it. This does not just include large appliances like refrigerators and washers, but also other smaller devices such as computers, phones and televisions sets. Because of this, reducing our use of appliances and equipment will likely not prove to be effective, as it would hinder our society in obviously negative ways. Therefore, improving energy efficiency of these devices while simultaneously switching to low-carbon sources of energy production is the most effective strategy. In 2019, energy efficiency in household appliances saved 281 TWh of energy![4] One of the most important areas for efficiency will be in heating and cooling equipment, as they also comprise the largest demand of energy consumption.

Heating and Cooling. Being able to control the climate inside of a building is a luxury that historically has only belonged to those who were the wealthiest in the world, and even then their capabilities were severely limited. The advent of electric air-conditioning and heating greatly expanded this privilege to markets in more developed countries such as Europe, North America, and Australia, but access to this remained outside of the grasp of many developing countries for a long time. China has been rapidly expanding its use of cooling units, and the same can be expected from India and Africa in the coming decades. In 2016, 60% of Chinese households had air-conditioning,[5] and Chinese markets consumed over 39% of global cooling equipment sales in 2019.[6] Today, industrializing African countries are now also expected to dramatically expand their access to heating and cooling (although obviously cooling is much more of a concern across most parts of the continent than heating is). All in all, electricity demand in buildings could increase by as much as 50% globally just by 2030 (compared to 2019) without "major efficiency improvements." [7] Heating sources primarily use fossil fuels, and this will have to transition dramatically to the use of electric equipment powered by renewables and heat pumps. (Geothermal and solar can also be used for direct heating in areas where it is applicable.)

The efficiency of refrigerants and cooling units should also be greatly prioritized because they use hydrofluorocarbons (HFCs)—which are extremely potent GHGs. The Montreal Protocol (the same protocol used to repair the Ozone Hole through the banning of chlorofluorocarbons, CFCs, from use in cooling and refrigerants) has now expanded its focus to include HFCs and hopes to reduce HFC consumption by over 80% before 2050.[8] If not properly addressed, HFC emissions could increase twentyfold in the next several decades, greatly increasing net GHG emissions.

Insulation. To ensure that improved heating and cooling measures are coupled with more efficient appliances and equipment, proper insulation of buildings is crucial. Without proper insulation across the board, large amounts of efficiency improvements will be lost. Some of the most impactful things you can do to make sure your house is properly equipped are to insulate your attic (this may sound monotonous, but it is actually a very big deal) and increasing the air sealing of your home around your windows and doors. Obviously, it's not always up to the resident, and in places where a landlord owns the property, it will be imperative that these factors are addressed in a cost-efficient and timely manner by the landlord. Because apartments within a complex tend to be identical (or at least very similar), these upgrades tend to cost less and take up less time per building, therefore making these updates even more easily implemented than on individual houses.

Summary on the Buildings Sector

In conclusion, buildings need to be updated dramatically. Some of the biggest things we can do is update efficiency standards and make sure these are met when creating new buildings and update existing buildings as much as possible. In most circumstances, tearing down buildings to start from scratch because of efficiency standards is unlikely to help as there are a host of emissions related with new materials and construction (not to mention concerns of displacement and other social impacts), and it simply does not make economic sense. What new buildings are built should be designed with energy efficiency in mind and should be built to last, as well as capable of supporting solar panels on the roofs if they are not installed by default. Failing to create sturdy, permanent, energy efficient structures could result in little to no emissions reductions.

Chapter 16

Helping Mother Nature
Carbon Capture, a Future Solution

The lack of realistic progress made to address the climate crisis has taken the world to the point where simply reducing emissions is no longer sufficient. In order to appropriately address our warming climate, there is a rapidly increasing consensus that actively removing GHGs (particularly CO_2) that have already been emitted back out of the atmosphere will be critically important in coming decades. This is called "Carbon Capture and Sequestration" (CCS). There are two schools of thought for solutions behind this: natural and technological. Natural CCS techniques involve practices such as afforestation (the planting of a new forest) and reforestation (the re-planting of a forest that was cut down or otherwise destroyed—almost always by human activities). Technological CCS solutions use methods such as chemical processing and separation to take CO_2 directly out of the air and then use it for other purposes—such as for use in enhanced oil recovery, where the CO_2 makes oil easier to extract from underground, and growing plants indoors (cannabis, for example, has a surprisingly high carbon footprint in part because of this)—or simply storing it for long-term. As the scales are somewhat unbalanced in discussions of these strategies, the difference between these two distinctions in methodology will be the main focus of this chapter.* One major downside to CCS techniques to keep in mind this chapter is that they almost exclusively target CO_2 and ignore other GHGs.

---

* Most often, natural CCS is seen as much more highly favored by the public and environmentalists. While this is not frowned upon, there is increasing data from the scientific community that this approach is either impractical or insufficient by itself and therefore is beginning to place much more attention on technological solutions.

Natural Carbon Capture

The most popular way that we go about practicing natural CCS is through tree planting. Among pretty much everyone, the large-scale planting of trees is welcomed with open arms. It's beautiful in a simplistic way. Anyone can go buy a tree sapling and plant it anytime and anywhere they want to, and it is something that communities and organizations across the globe can collaborate on to have an impact. Additionally, since lumber companies—in many parts of the world—are now responsible for replanting the trees that they cut down, the effects of deforestation are (to an extent) being mitigated. Regrettably, however, planting trees is not always the most practical way to fight climate change. This is mainly because planting forests simply takes up too much space. If we wanted to completely negate current global emissions through the use of forests, we would need to cover a simply impractical amount of land area. Planting trees is a good way to minimize the harmful effects of climate change, but not a great way to absorb GHGs.

In 2018, an estimated 37.1 Gt of anthropogenic CO_2 was released into the atmosphere.[1] The same year, energy-related CO_2 emissions potentially accounted for 33.5 Gt of this.[2] In 2010, the world produced approximately 49 Gt of anthropogenic CO_2 equivalent GHG emissions[3] (today this figure is closer to 51 Gt), and energy-related emissions accounted for 76% of this.[4] * Despite the massive amounts of emissions, the natural environment has historically absorbed large amounts of these. For example, the world's oceans absorb about 31% of CO_2 emissions, and from 1994-2007, they absorbed an estimated 34 Gt of CO_2.[5] Between 1800 and 1994, they have absorbed approximately 118 Gt.[6]

More to the point of this chapter, trees also remove large amounts of carbon from the atmosphere. While they can easily be planted on large scale, there are several limitations to this, and the effect to which it can help is highly debated. As we will see, the large-scale planting of

---

* Energy-related emissions in this case includes emissions from the following sectors: Electricity & heat production (25%), Industry (21%), Transport (14%), Buildings (6.4%), and Other energy (9.6%).

trees could be much more complicated than is commonly perceived. Additionally, while this chapter will focus on tree planting, there are also proposals for CCS through the use of other plants like kelp and algae, though there is debate about the practicality of these as well.

Land and Cost Requirements. To offset the massive amounts of CO_2 being emitted, an IPCC special report noted that "…planting 1 billion hectares of trees would be necessary to prevent global temperatures from increasing by 1.5 degrees Celsius by 2050." [7,8] There are various studies that explore the practicality of this. One report's co-author stated that his study found that afforestation is "by far—by thousands of times—the cheapest climate change solution" and that this may entail adding 1-1.5 *trillion* trees.[9] Another found that planting trees to this scale is "among the most effective strategies for climate change mitigation." [10] Yet another found that "Natural climate solutions can provide 37% of cost-effective CO_2 mitigation needed through 2030 for a [greater than] 66% chance of holding warming to below 2°C." [11]

However, there is *significant* debate about the effectiveness, ethicality, and cost of these proposals.[12,13] As there are approximately 3.04 trillion trees on the planet currently,[14] this could mean increasing the number of trees on the planet by a staggering 30-50%. Granted, humans have vastly reduced the tree coverage on the planet and there is plenty of room to be taken back, but the scale of this project would be prodigious. In all reality, without a massive change in the way humans occupy land (primarily from agriculture), the implementation of a project this size will only become increasingly difficult. Energy requirements (and therefore, most likely, GHG emissions) are expected to continue rising along with population over the next century. As these requirements grow, the massive expansion of onshore wind and solar energy could also potentially have an impact on available space for these projects. Even further, as underdeveloped populations experience rapidly growing economies, their land usage will increase—even as they move to urban areas—because of increased infrastructure, industrial-scale agriculture, and growing industrial sectors. Other industries such as mining—which is also poised to grow massively due to the upcoming "green" industrial revolution—also require large areas of land.

Additionally, the world may already be slowly greening. One study published in the journal *Nature* found that globally, tree coverage had increased by roughly the size of Texas and Alaska combined between 1982 and 2016.[15] Much of the reasoning given for this was due to reforestation and global warming (from trees being able to expand into what was previously colder climates). Higher concentrations of CO_2 has also likely helped spur tree growth, though it is imperative to understand that too much CO_2 can have the opposite effect and actually inhibit their growth (the point at which we are already beginning to reach with some crops such as rice).[16,17] Just as humans can get water poisoning from drinking too much water, plants can get hurt when ingesting too much CO_2.

Biodiversity. Tree planting could have significant biodiversity implications. Mono-cropping (the planting of a single species of tree) is a common strategy for tree planting, but this is obviously not how nature is in reality, and a single species of tree being widely planted could therefore potentially have large negative consequences on existing ecology. For example, one insect might thrive under this type of tree, allowing for the proliferation of said species and may therefore cause damage to the stability of the local ecosystem. Additionally, if a foreign tree species is introduced into a new ecosystem, it could become invasive—leading to other ecological consequences—or be quickly outcompeted by native ecology, resulting in a loss of time and money with no real carbon sequestration benefits. Because of these issues, extensive studying of a site should be conducted before implementing afforestation techniques.[18]

These problems are only compounded when considering that trees that absorb the most CO_2 (these are also generally the trees that grow the fastest) would be preferable, although they will not be applicable to every site. The species and quantity of trees will need to be specific to each site to allow for minimal negative ecological impacts. Compared to afforestation, reforestation may not be as problematic due the fact that the species which existed in the spot prior to being cut down are known, and would help the previously damaged ecosystem recover from any previous disturbances.

Climate Change Considerations. As the climate changes, tree coverage and habitats will be massively affected. A wildfire, for example, could have the potential to destroy massive swaths of planted forests very quickly, negating the carbon captured from the site. As droughts become more frequent and desertification expands, areas that are suitable for tree growth today may become less effective in the future (though the opposite is also true).

Furthermore, as some diseases begin to spread easier, monocrop tree sites could be at high risk. In a worst-case scenario, one disease or insect plague could destroy an entire site. Site-specific trees of a variety of species could be significantly more resilient to climatic changes than monocrops and may therefore be worth the increased costs. A massive increase in the amount of global tree coverage would also use large amounts of rainwater, which could negatively affect crop yields and water security (especially due to increased water shortages that may come from global warming). Finally, some forests may have a net warming potential instead of a cooling potential. The planting of a forest in certain locations may reduce the land's albedo, allowing more heat to be absorbed by land instead of being reflected back out into space, which in some cases could actually warm the planet more than the CO_2 that is being removed.

Outside of the potential that trees have to mitigate CO_2, they may also have large potential for mitigating the impacts being felt by climate change. This includes flood protection, climate stabilization, and slowing desertification. This is the premise behind Africa's "Great Green Wall" (China also had a project by the same name). This super-project aims to grow a wall of trees across the entire width of Northern Africa, which will help to prevent/slow the expansion of the Sahara Desert and regional droughts, while also improving prosperity. The projects website defines itself as "…a global symbol for humanity overcoming its biggest threat – our rapidly degrading environment." [19] Mangrove tree planting is also a popular solution, which helps to reduce flood surges in coastal regions. An added benefit of mangroves is that they permanently sequester carbon far below the surface, so when they die, the amount of carbon released is far lower than most trees. Something similar is seen with aquatic crops such as kelp.

In urban areas, tree planting can have substantial positive impacts on communities by helping to clean the air, reducing the urban-heat island effect (in premise, urban areas are much hotter than natural/undeveloped land because of man-made structures) and at the very least providing noticeable benefits to our mental health.

Technological Carbon Capture

Technological solutions for CCS look at directly taking CO_2 out of the air and storing it somewhere else (typically underground) or utilizing the carbon for industrial activities. Many of these solutions have several advantages over natural sequestration. First, technological solutions allow for the utilization of captured carbon—Carbon Capture Utilization & Storage (CCUS)—as well as storage. This gives the added advantage of possible real-world business models that can help to scale and develop the technology further, while storage could make the technology carbon-negative. Ideally, this will accelerate CCS technology to the point where the "utilization" part is no longer needed for economic benefits (due to government carbon-credits for carbon-negative solutions). In the SDS, CCUS removes around 9 Gt CO_2/y by 2050 between various sectors (namely industry, power, and energy).[20] However, today CCUS removes essentially no CO_2 because it is still a new and expensive technology. Unfortunately, all forms of CCUS focus on CO_2, and there are not yet any major ways to reduce emissions from HFCs, CFCs and methane.

CCS at the Source of Emission. CCS has massive potential to mitigate emissions at the source. CCS systems can be attached to fossil fuel plants or industrial facilities to capture most or all of the produced carbon before it even escapes into the atmosphere. This could potentially allow for a slightly decelerated (and therefore more natural) reduction of fossil fuels while also reducing GHG emissions. In 2020, there were 21 sizeable CCUS projects in operation, 11 of which were in the United States.[21,22] There are also many more that are under construction around the world. To comply with the SDS, this figure will need to grow exponentially. CCUS is still far behind where it needs to be. In 2018, existing capacity of CCUS was just 2.4 Mt CO_2/y in the power generation sector—less than 1% of what it needs to be in 2030 under the SDS.[23]

Direct Air Capture. Outside of capturing emissions at the source, CO_2 can also be captured straight from the air; this is called Direct Air Capture (DAC). DAC will be critically important for reducing CO_2 emissions. Since emissions from sectors such as agriculture and transportation cannot be captured at the source, DAC can help mitigate them by sucking CO_2 right out of the air anywhere on the planet. Because the emissions are not captured at the source, DAC is less efficient and requires larger amounts of power than the previous techniques discussed. This combination generally makes it more expensive than CCS at the source of emission.

Once captured from the air, DAC stores the captured CO_2 underground. There are multiple ways that it can be effectively stored this way, but one of the most popular is by pumping it in with basalt, where it will mineralize, becoming a solid, and then stays there indefinitely without fears of leakage.

Summary on Carbon Capture

It's really the cost of these different technologies that are hindering them as of right now. As costs decrease, they will become more and more viable. CCS is going to be needed in the fight against global warming as we change over to clean energy, and it is especially important with the noticeable lack of effort globally to reduce emissions. As put by the United Nations Economic Commission for Europe,

> "If the world is to succeed in constraining CO_2 emissions to levels consistent with a less than 2°C rise in global temperatures, then Carbon Capture and Storage (CCS) will need to contribute about one-sixth of needed CO_2 emission reductions in 2050, and [14%] of the cumulative emissions reductions between 2015 and 2050 compared to a business-as-usual approach. It is the only technology option other than energy efficiency and shifting the primary energy mix to lower carbon fuels that can deliver net emissions reductions

at the required scale. The IPCC Fifth Assessment Synthesis Report estimated that without CCS the cost of climate mitigation would increase by 138%." [24]

Unfortunately, CCS technology is just not developed enough to the point where it can make a large difference in our emissions. If the technology had been researched more several decades ago it could already be in wide use, but its usefulness is probably similar to nuclear fusion. It may be able to help at large scale in the second half of the century, but currently it is not ready. To catch up to where the technology is needed, massive government investments will be needed.

Section IV:

Is All of This Necessary?

Chapter 17

The Need for Action
Introduction to the Crisis at Hand

Do we really need to change our ways so dramatically? Could climate change actually be so horrible that we need to do all of this? Are fossil fuels truly so bad that we need to completely eliminate them? Are there better ways to do this? Wouldn't it just be easier to adapt to climate change rather than completely uproot the way society operates?

These are the questions this final section of the book aims to address. They are all distinctly important. As climate change affects the whole world, it must be all-hands-on-deck. If it would be determined to rather focus on adaptation, then the opposite would be true. Some countries, most notably Canada and Russia, could benefit from global warming by large areas of their land being made arable from melting permafrost. As the permafrost melts, more and more people will be able to settle, build, and grow crops in these regions. Because of this, several have proposed that we should welcome global warming. Svante Arrhenius, who you might remember from the introduction to this book, used his early research on GHG emissions to argue that we should try to warm Earth in order to provide more food and stimulate plant growth. He envisioned using GHGs to make Earth's climate "more equable." [1] These same arguments continue to be put forth today, even though the scientific literature is clear that climate change will have a net-negative impact on the planet.

This section may be the most important content in this entire book simply because in order to understand what we need to do, we first need to be able to agree that we need to do something in the first place. The US is increasingly unique in this sense. The vast majority of citizens and countries around the world agree that anthropogenic climate-change is happening and will negatively affect the world, but many in the US continue to debate this issue and sew doubts of if the science is settled, accurate, or if there are malicious intentions behind the transition. This is not my opinion either. A 2018 *PEW* survey found that the US was among the top countries whose citizens do not consider climate change a threat,[2] and another survey found Indonesia and the US to be the most in denial of climate change or that humans are responsible for it.[3] A Yale national study estimated that only 72% of American adults believe climate change is happening (12% saying no and the rest undecided) as recently as 2020.[4]

This is very worrisome when considering that the *vast* majority of climate scientists believe that climate change is happening, manmade, and will have severe consequences.

Scientific Consensus

I have done my best to keep personal opinions out of this book and stick to available data. However, I will make a single remark. The idea of the scientific consensus on man-made climate change "not being settled" is nothing but preposterous.

"Multiple studies published in peer-reviewed scientific journals[5] show that 97 percent or more of actively publishing climate scientists agree: Climate-warming trends over the past century are extremely likely due to human activities. In addition, most of the leading scientific organizations worldwide have issued public statements endorsing this position." [6] In 2009, 18 separate American associations signed a letter urging for climate action.[7] This included the renowned *American Association for the Advancement of Science*, the *American Chemical Society*, the *American Geophysical Union*, the *American Institute of Biological Sciences*, the *American Meteorological Society*, the *American Society of Agronomy*, the *American Society of Plant Biologists*, the *American Statistical Association*, the *Association of Ecosystem Research Centers*, the *Botanical Society of America*, the *Crop Science Society of America*, the *Ecological Society of America*, the *Natural Science Collections Alliance*, the *Organization of Biological Field Stations*, the *Society for Industrial and Applied Mathematics*, the *Society of Systematic Biologists*, the *Soil Science Society of America*, and the *University Corporation for Atmospheric Research*.

Additionally, government organizations who are relevant to this topic have long held the position that climate change is real, even when the Commander in Chief is staunchly opposed to recognizing climate change, in which case taking a hard stance against said President's opinions may result in budget cuts or political attacks (as was seen under President Trump). The US *Fourth National Climate Assessment*, compiled by various government organizations, demonstrates this.[8] The report—which firmly establishes climate change will be of direct threat to many Americans—was compiled through collaboration of 13 distinct US federal departments and agencies: The *Department of Agriculture*, the *Department of Commerce*, the *Department of Defense*, the *Department of Energy*, the *Department of Health & Human Services*, the *Department of the Interior*, the *Department of State*, the *Department of Transportation*, the *Environmental Protection Agency*, the *National Aeronautics & Space Administration* (NASA), the *National Science Foundation*, the *Smithsonian Institution*, and the *US Agency for International Development*.[9] (The *National Oceanic and Atmospheric Administration* (NOAA), the *US Geological Survey* (USGS), and various other departments also play massive roles inside of the US government in relation to climate change.)

There are countless private organizations that confirm this viewpoint such as *Scientific American Magazine* (who, for the first time in the magazines 175 year history, endorsed Joe Biden for president over Donald Trump in 2020, largely due to Trumps stance on climate change).[10] Virtually all serious media outlets confirm this even further.

Even Fox News, who, while regularly promoting climate skepticism on TV, has held a conflicting position on climate change in online publications, varying widely between reporters. Some saying it is not real, other saying it's real but we shouldn't worry, others still saying it's a problem but there is nothing we can do, and yet more saying that we can and should fix the problem. This promotes the incorrect perspective that the science is not settled. This is not by coincidence either. Like Big Tobacco last century, Big Oil actively seeks to spread the false notion that climate science is not settled through tactics just like this. If you can make it seem like there is true debate going on through the mainstream media, most will believe that there really isn't a consensus.

Not to mention all serious environmental organizations such as *Greenpeace*, the *Sierra Club*, *Environmental Progress*, *350.org*, the *Environmental Defense Fund*, and *The Nature Conservancy*, just to name a few. Various colleges consistently put out publications pertaining to climate change and the need for action, including *Yale*, *Stanford*,

Harvard, Princeton, MIT, the *University of California,* and countless others. This is all to say nothing about the overwhelming amount of scientific data coming through various journals and publications, the volume of which could induce vertigo.

As far as international organizations go, this consensus is only amplified. The *United Nations* holds a strong opinion on this subject. The richness of study data and evaluations coming from within the UN about this topic is impressive, to say the least, and the frequency with which more of this data is released is unprecedented. The established *Intergovernmental Panel On Climate Change* (IPCC) is the obvious subject here, where thousands of scientists from across the globe have collaborated for several decades and have provided clear and irrefutable evidence of man-made climatic changes. Along with the IPCC, the *Food and Agriculture Organization,* the *International Monetary Fund,* the *International Maritime Organization,* the *United Nations Educational, Scientific and Cultural Organization* (UNESCO), the *World Health Organization,* the *World Meteorological Organization,* the *World Bank,* as well the *United Nations Environmental Programme* (UNEP), the *United Nations Framework Convention on Climate Change* (UNFCCC), the *International Energy Agency,* the *International Atomic Energy Agency,* and others consistently publish information discussing impacts of climate change on their respective focuses. Essentially, any serious organization around the world holds the position that climate change is ongoing, has been caused by human activities, and will have major rammifications.[11]

Perhaps it is an institutional problem. What do individuals have to say? There are many high-profile individuals fighting against climate denial, many of whom are climate scientists. There are thousands of leading scientists (a number that outweighs the amount of scientific climate deniers by several orders of magnitude) as well as most politicians around the world (even essentially all of the right-leaning parties in countries outside of the US acknowledge climate change as a serious threat) and what is likely millions of activists. Two of the most prominent figureheads from this are former Vice President Al Gore and the British broadcaster David Attenborough. Other prominent figures include Bill McKibben, Michael Mann, US Representative Alexandria Ocasio-Cortez, Robert Pollen, Bill Gates, Noam Chomsky, Elizabeth Kolbert, James Hansen, Jane Goodall, Leonardo DiCaprio, Naomi

Klein, and many others.

Additionally, there has been a war of letter signing going back and forth in the US for several years now. One year, a few thousand "experts" will sign a letter attesting that climate change is of no concern, and the next another few thousand will sign on the opposite. Both sides have the issue that these letters are often simply not good methodology. If anyone can sign it, it becomes irrelevant. A random person's opinion on this does not really matter, but what matters is the opinion of the scientists who work *specifically in climate research*. This makes understanding individuals' opinions tricky to understand, though various peer-review studies examining scientific publications have found the overwhelming majority of peer-revied studies (~97%) to be of the assessment that climate change is real, and it is manmade.

To take just one high-level example: The "Global Warming Petition Project" (also known as the Oregon Petition) acquired over 31,000 signatures from "American Scientists" [12] who attested to the non-existence of climate change and was flaunted around, but the credibility of this was "easily debunked." [13] Of the signatures, fewer than 40 were in the field of climate science, and many were not scientists at all or did not provide any information on their background. The only restricting requirement was a bachelor's degree. What if someone made a petition to give everyone $1 million a year and got thousands of signatures, but the vast majority of signees had degrees in art, instead of being economists? This is the same situation seen with the Oregon Petition. Admittedly, there were a number of impressive, high-profile signees; fascinatingly, Charles Darwin managed to rise out of his grave—where he was placed in 1882—to sign the petition. Even characters from *Star Wars* and (people impersonating) the Spice Girls signed their names on the petition. Very convincing indeed.

Another instance that is commonly pointed out was the so-called "Climate Gate" incident where hackers who gained access to emails being exchanged between scientists in the lead-up to the 2009 UN Climate Change Summit in Copenhagen. These emails were then released to the public, and various sources pointed out that the way some of these emails discussed climate change proved a massive conspiracy. Some phrases like using a "trick" to "hide the decline" and "The fact is that we can't account for the lack of warming at the moment and it is a travesty that we can't" were among those touted as this proof. In reality, these phrases were taken severely out of context and manipulated to create false pretenses.

The "hide the decline" phrase was used when discussing a specific strategy relating to tree-ring readings that were known to be wrong. Essentially, sometimes tree rings from certain regions would show a decline in temperature even though it was known through observation that this was not the case. (The specific issue here is called the tree-ring divergence problem and is well studied and accounted for.) In the latter case, the scientists were not admitting they could not account for a lack of global warming. As the target of this controversy stated, "It is amazing to see this particular quote lambasted so often. It stems from a paper I published this year bemoaning our inability to effectively monitor the energy flows associated with short-term climate variability. It is quite clear from the paper that I was not questioning the link between anthropogenic greenhouse gas emissions and warming, [and possibly] even suggesting that recent temperatures are unusual in the context of short-term natural variability." [14]

What about those that promote climate denial? There are few specific organizations that promote climate change denial that are worth bringing to attention. These include the *Heartland Institute* (which received $4.5 million from Koch foundations from 1997-2011 and $780,000 from ExxonMobil from 2001-2012), the *Institute for Energy Research* (IER, which has received funding from the Koch Brothers and ExxonMobil), the *Cato Institute* (which was co-founded by Charles Koch, and has received millions of dollars from Koch foundations), *Americans for Prosperity* (to whom "Koch foundations donated $3,609,281…from 2007-2011"), and the *American Enterprise Institute* (which received over $3.6 million from ExxonMobil between 1998 and 2012, and over $1 million from, you guessed it, Koch foundations from 2004-2011).[15] Other notable mentions include the *American Legislative Exchange Council*, the *Beacon Hill Institute*, the *Competitive Enterprise Institute*, the *Heritage Foundation*, and the *Manhattan Institute for Policy Research*.

This is not a new occurrence or unique to climate change either. In the landmark novel *Merchants of Doubt*, Naomi Oreskes and Erik M. Conway discuss this subject matter in depth.[16] Throughout this reading, the authors make the same observation over and over under different situations. The similarities between corporate denial and the spreading of disinformation about the negative health effects from tobacco smoke, acidic rain, climate change, and others are all shown in frightening detail. The difference is while today nobody would argue that cigarettes can be harmful, many will still argue that climate change is insignificant or not harmful, despite the plentitude of data that backs it (as is shown later).

It is even more frightening when considering that, as *Merchants of Doubt* lays out in excruciating detail, the exact same individuals who were responsible for the denial of tobacco smoke, acid rain, harmful pesticides, climate change, and other subjects were often involved in the denial campaign of not just one, but several of these issues. Despite being proved that tobacco smoke is indeed harmful, the same people who advertised it wasn't harmful are now a major driving force behind the climate change denial movement, getting rich by selling doubt as a product.

What about arguments used against climate change? Many of the arguments against climate change that are put forth are red herrings: The temperature and CO_2 concentrations have been higher in the past! The climate changes naturally! The amount of CO_2 in the atmosphere is incredibly small compared to water vapor, which is also a GHG! Without the greenhouse effect, the Earth would be freezing cold!

Yes, CO_2 concentrations and temperatures have been higher in the past (the last time atmospheric CO_2 levels were this high was likely several million years ago), but humans and other species around the planet are not adapted to these temperatures, not to mention are having a difficult time acclimating to the rapid pace of the change. This argument also completely ignores the fact that when CO_2 levels were higher millions of years ago, the energy from the Sun that Earth received was less than today, meaning that in reality the amount of energy the planet absorbed is relatively similar in both of these time periods: i.e., even though CO_2 levels were higher the temperature of the planet was not that much warmer because the sun was weaker. Yes, the climate changes naturally, but the change happening now is clearly far from natural due to its pace. Yes, water vapor is a GHG and there is far more of it in the atmosphere than CO_2, but water vapor also has a cooling effect and CO_2 is a much more potent GHG (per volume) than water vapor—to say nothing of other GHGs such as methane, CFCs, and HFCs (the amount of water vapor in the atmosphere is therefore largely driven by other GHGs). Yes, without the greenhouse effect, the Earth would be cold and likely not suitable for life as we know it, but we do not want it to be smoldering hot either, as Venus is.

There are also arguments against climate change that are just plain wrong. Volcanoes, for example, are often pointed out as producing much more CO_2 than humans. This is completely false. "All studies to date of global volcanic carbon dioxide emissions indicate that present-day subaerial and submarine volcanoes release less than a percent of the carbon dioxide released currently by human activities." It would take approximately 3,500 Mount St. Helens-equivalent eruptions to equal 2010s anthropogenic CO_2 emissions.[17] (Additionally, volcanic eruptions often have a strong global cooling effect, as has been seen numerous times in history, because the sulfur dioxide that they release makes the atmosphere more reflective, in turn causing the planet to absorb less energy.) Others argue that it isn't happening at all, which has been proved wrong by so much empirical data that, frankly, I don't even know what to use for a citation here.

This debate may not have been as significant of an issue if the country most in denial was a smaller, less influential country with a much lower carbon footprint. However, the US is a world leader. Hesitancy from the US on this front creates uncertainty in the economic and intergovernmental policies that attempt to make a coordinated effort to solve a global issue much more difficult. Of course, the US is not the only country to blame. Australia, Indonesia, and the UK are also strong climate denial hot spots, even though each of these countries are already experiencing severe climate change impacts.

In 2014 then UN Attorney General Ban Ki-moon said, "Climate change is coming much much faster [than anticipated]. We have seen such extraordinary extreme weather patterns. If you consider this vastness of this universe, this planet earth is just a small boat. If this boat is sinking then I think that we will all have to sink together." To prevent our collective lifeboat from sinking, we first have to acknowledge that we are already taking water.

The next two chapters will focus first on why (regardless of climate change) we should move away from fossil fuels, and secondly what impacts we could expect to see if we ignore climate change.

Chapter 18

Habitual Suffering
Fossil Fuels, an Outdated Energy Source

Fossil fuels have done more good for human civilization than anything else since the invention of fire, language, and agriculture. Without them, the world would not have been able to industrialize and rapidly grow (though obviously this rapid growth has in-part caused the problem we are now in), which has brought our species previously unfathomable prosperity, democracy, human rights, and even the poorest people on the planet are relatively better suited to the common person a couple hundred years ago. Historically, centuries could go by with very little economic or social progress, making our new world growth unprecedented. Easily accessible and cheap electricity would not have been available essentially anywhere (without steel and concrete made from a fossil-fuel based system, large-scale hydro, solar, wind, geothermal, or nuclear power would have been impossible). Horse and buggies would still be a major source of transportation. Boats would still use sails, severely limiting global trade. Widely available steel, concrete, electronics, and many other commonly used things would likely not exist. Early trains powered by coal would not have existed, making long-distance travel and trade over land much more difficult and limiting the pace of development (it is more likely that without trains the west coast of the US would only as of recently have been extensively settled). And there are infinitely more examples of how fossil fuels changed the world for good. Most people, certainly, would agree that the world is far better with the existence of fossil fuels than it would have been without them, even when considering the problems of air pollution and man-made climate change.

Despite this, we also know that fossil fuels (particularly coal) are also horrible. Coal has caused very major health impacts in developing countries for hundreds (and in the case of ancient China, thousands) of years. Scientists have been warning about the damaging effects on climate change for decades, and yet very little has changed. The argument for a long time was that there was no conclusive evidence that fossil fuels will contribute to climate change, but that argument does not stand anymore as there is an irrefutable consensus and amount of evidence (see chapters 17 and 19). Even the fossil fuel companies have a complete understanding of how much their business damages the environment, but nobody is going to promote their own demise. Upton Sinclair once famously said, "It is difficult to get a man to understand something when his salary depends upon his not understanding of it." This is perhaps why man-made climate change has been so feverishly contested over the past several decades, especially in places where fossil fuels are especially integral in the economy. (The US both produces and uses massive amounts of oil, coal and natural gas, Australia is responsible for about 1/3rd of global coal exports, and Indonesia also exports massive amount of coal, making the fact that these countries are hot spots for climate denial relatively unsurprising.)

In 2015 it was revealed that ExxonMobil had known about the implications that its business would have on the global climate for a long time,[1] which then led to its being sued for misleading future investors (Exxon won the lawsuit in 2019) and the #Exxonknew trend along with a further deep public distrust of not only ExxonMobil but all fossil fuel corporations.[2] What we know now is that the oil giant had indeed spent massive amounts of time, money, and resources investigating climate change in the past. Once it was revealed by its own scientists that the burning of fossil fuels would create destructive global warming, the company began large disinformation campaigns and created organizations—such as the Global Climate Coalition—that were designed to sew doubt into the public perceptions on the science of climate change. As early as 1977, Exxon was aware of how its business could impact climate change.[3,4] In July of 1977, a senior scientist at Exxon told high-level officials in the company, "In the first place, there is general scientific agreement that the most likely manner in which mankind is influencing the global climate is through carbon dioxide release from the burning of fossil fuels." [5] Exxon would then go on to dismantle its climate research program, and its former scientists could not speak out due to the signing of a Non-Disclosure Agreement with the company. This is identical to what happened with tobacco smoke. Once Big Tobacco's scientists realized how harmful it was, they were silenced, and the companies went on to fund gargantuan disinformation campaigns.

Even though renewables are beginning to make a positive impact, in 2020 60.3% of US electricity was still generated from fossil fuels.[6] Furthermore, natural gas power production has created extremely competitive prices for electricity and heating, and therefore is growing in popularity, though the total share of fossil fuel consumption is reducing.

This chapter is included in the final section of this book and not earlier because while fossil fuels are a way for producing energy, they are not a solution. The link between fossil fuel GHG emissions and climate change are well known and observed, so this chapter will not focus on this factor. Instead, this chapter will make the following argument:

Even if man-made climate change was not occurring, we should still transition off fossil fuels. They are inefficient, expensive, and have major health and environmental impacts outside of the context of climate change.

Health and Environmental Impacts

Fossil fuels, particularly coal, are extremely deadly. Coal mining accidents kill an inhumane amount of people every year. On average, more people every year die in coal mines than have ever died directly from nuclear, solar, and wind power, combined. Although coal mining is very deadly everywhere, China in particular has a particularly horrible history:

> "China's total death toll from coal mining to 2008 averaged well over 4000 per year – official figures give 5,300 in 2000, 5,670 in 2001 and 6,995 in 2003, 6,027 in 2004, about 6,000 in 2005, 4,746 in 2006, 3,786 in 2007, 3,210 in 2008, 2,631 in 2009, 2,433 in 2010, 1,973 in 2011 and 'over 1,300' in 2012. ... However, the picture is improving: in the 1950s the annual death toll in world coal mines was 70,000, in the 1980s it was 40,000 and 1990s it was 10,000. Additionally; Ukraine's coal mine death toll has been over two hundred per year (eg. 1999: 274, 1998: 360, 1995: 339, 1992: 459). In Australia 281

> coal miners have been killed in 18 major disasters since 1902, and
> there have been 112 deaths in [New South Wales] mines since 1979,
> though the Australian coal mining industry is considered the safest
> in the world. The USA, the world's second-biggest producer, recorded
> 48 coal mining deaths in 2010." [7]

This also doesn't include the incalculable deaths that have resulted from coal mining in the past several hundred years, where coal mines were essentially graveyards, and employed large amounts of child labor. Although the signs are showing that safety is improving, it seems inhumane that we could even have allowed this to occur for so long. The nature of coal mining is simply inherently dangerous, and regardless of how safe the industry becomes, there will always be a large risk looming overhead (though, granted, many use the exact same argument against nuclear power and the intermittency of renewables). There will likely always be accidents that kill people in this industry, especially in less developed or developing countries where safety is not as much of a concern and child labor continues to persist.

The largest coal mine incident so far this century occurred in 2014 in Soma, Turkey. An intense mine fire began, and, as the fire burned, parts of the mine collapsed and large amounts of toxic fumes were released. In all, 301 people were reported killed in the incident and another 80 suffered serious injuries.[8] This is to say nothing of the immense coal seam fires that continue to burn around the world. The Centralia, Pennsylvania coal fire that began in 1962—which resulted in the slow destruction and abandonment of the entire town built around the coal mine and a stretch of highway—is expected to continue burning for nearly 250 years. In Zwickau, Germany, a coal seam fire that reportedly began in the third quarter of the 15th century was only stopped in 1860—nearly 400 years later. Also in Germany, the "Burning Mountain" located in Dudweiler that has been burning since 1668 continues to this day.* In 2002 a coal fire that has been burning since the late 19th century in New Castle, Colorado (strangely, also commonly

---

* This name is more commonly seen in German as the "Brennender Berg" and there is a natural coal seam fire that also goes by "Burning Mountain" in Australia.

referred to as "Burning Mountain") ignited a surface fire that destroyed twenty-nine homes and burned over 12,000 acres of land. And these are just a sampling of coal seam incidents.

Although coal is certainly the worst, oil and natural gas are not free from incidents either. To name a couple, in 1998 a Nigerian oil pipeline that served to move oil from a refinery to the towns of Warri and Kaduna exploded and caused a major fire. Nearly 1,100 deaths were resulted, and over 300 of the bodies were so badly burned they were deemed unidentifiable and were buried in mass graves.[9] A natural gas explosion in 2003 killed 191 people in China, which also resulted in the evacuation of another 41,000 people.[10] Oil spills have not only caused death of workers and major economic losses, but also serious environmental and social problems. For example, after the Deepwater Horizon oil spill in the Gulf Coast, many fishermen's livelihoods were uprooted for extended periods of time and the regional tourism suffered significant losses when compared to expected revenue otherwise.

Unfortunately, these incidents are just scratching the surface of the true dangers of fossil fuels. This is because despite everything previously mentioned, fossil fuels kill silently. The emissions created by fossil fuels obviously produce CO_2—which is bad enough on its own—but they also release things like sulfur dioxide and nitrogen dioxide (which are responsible for acid rain), mercury, carbon monoxide, methane, radon, and various other elements—along with radionuclides such as uranium or thorium— that can be dug up with the coal. The end-result of these emissions represent what is often referred to as "Particulate Matter" (PM). PM is simply the mixture of solid particles and liquid droplets found in the air, as defined by the EPA.[11] This complex mixture includes both organic and inorganic particles, such as dust, pollen, soot, smoke, and liquid droplets. PM (specifically PM2.5)* is responsible for the negative health effects of general air pollution and also contributes to acid rain and smog.[12]

Air pollution (and thereby PM) has been strongly linked to conditions such as lung cancer, heart disease, aggravated asthma and

---

* PM2.5 refers to the size of the particulate. The "2.5" means the particulate is 2.5 micrometers (or smaller) in size. PM2.5 is where the hazards become serious, though other particulate sizes can also be dangerous.

emphysema,[13] but have also been linked to neurological disorders such as Alzheimer's, Parkinson's disease, and dementia.[14] Basically any type of negative health effects, aside from immediate physical harm, can be linked to PM2.5. Aside from fossil fuels, wildfires and generally hotter temperatures have also been releasing large amounts of particulates from within the ground; although fossil fuels have historically contributed to this the most (and likely will except for in the future occurrence of a cataclysmic event such as a super volcano eruption or major asteroid impact—in which case massive amounts of dust and matter would be kicked up into the atmosphere). Despite this, as global warming continues, wildfires and global temperatures will emit more and more hazardous materials into the air, exacerbating the health effects on both humans and wildlife.

A 2014 WHO report shows the staggering effects that air pollution has on human health. Although we can't see it and it seems far away, it directly impacts everyone and everything in a very serious way. The report states that, "…in 2012 around 7 million people died - one in eight of total global deaths – as a result of air pollution exposure. This finding more than doubles previous estimates and confirms that air pollution is now the world's largest single environmental health risk," [15] and another study in 2021 found global deaths from particulate pollution to be much higher than even this, estimating 10.2 million excess deaths from PM2.5 in 2012.[16]

The burning of fossil fuels for electricity and the powering of our vehicles is responsible for the vast majority of air pollution, meaning that they indirectly kill millions every year. Roughly 1.6 million people in China die prematurely every year from health conditions caused by air pollution—over 4,000 people every day.[17] A whopping 38% of the Chinese population lives in an area that would be deemed "unhealthy" by the US EPA.[18] Aside from particulates, things such as black lung— pneumoconiosis resulting from inhaling coal dust—have killed many coal miners and workers in the fossil fuel industry, which has had severe implications. "…you shouldn't have to sacrifice your life for your livelihood. But that's been the fate of more than 76,000 miners [in the US] who have died at least in part because of black lung since 1968." [19]

In the United States, the air quality has greatly improved in the past couple decades. During its industrial revolution, there were immense levels of air pollution harming the ecosystem and citizens alike, but the country has since implemented The Clean Air Act, which vastly helped to reduce air pollution. This included regulations on emissions from power plants and manufacturing, waste disposal reform, and emissions regulations on automobiles. Although the air in the US is generally considered to be at its best since the industrial revolution, in 2020 the American Lung Association noted in its annual *The State of the Air* report that "Nearly five in ten people—150 million Americans or approximately 45.8 percent of the population—live in counties with unhealthy ozone [which can be formed from fossil fuel use] or particle pollution. That represents an increase from the past three reports: it is higher than the 141.1 million in the 2019 report (covering 2015-2017), 133.9 million in the 2018 report (covering 2014-2016) and 125 million in the 2017 report (covering 2013-2015). ...This shows growing evidence that a changing climate is making it harder to protect human health." [20] In the US, California claimed the titles for the cities with the worst ozone pollution (Los Angeles), the worst year-round particle pollution (Bakersfield), and worst short-term particle pollution (Fresno-Madera-Hanford).[21] There are several reasons why California's air pollution is so bad, but the prime driver is due to warm weather and constant wildfires coupled by intense fossil fuel consumption (historically, California has had some of the United States' worst air pollution). As far as climate changes correlation to air pollution, *The State of the Air* report had clear points to make:

> "This shows growing evidence that a changing climate is making it harder to protect human health. All three years [2016-2018] ranked among the five hottest years in history, increasing high ozone days and widespread wildfires, putting millions more people at risk and adding challenges to the work cities are doing across the nation to clean up. Rollbacks of EPA cleanup rules and reduced Clean Air Act enforcement are further adding to these air quality challenges. This marks the fourth report in a row that worsening air quality threatened the health of more people, despite other protective measures being in place. Climate change clearly drives the conditions that increase these pollutants. The nation must do more to address climate

change and to protect communities from these growing risks to public health." [22]

As countries industrialize, their air quality drops rapidly, but then improves again as the state takes the threats of air pollution and general public health seriously. This explains why China's air (although notoriously bad) is improving over recent years. Today the twenty cities in the world with the worst air pollution are all in Asia, but fifteen of them are in India, who is now going through heavy industrialization.

While coal is the worst form of electricity production for both air pollution and climate change impacts, natural gas is certainly not perfect. Hydraulic-Fracturing (fracking) is the process used to extract natural gas and oil from below the surface. Fracking can be extremely volatile to both the environment and populations around them. The process of fracking has been shown under some circumstances to directly impact the health of people and livestock living nearby the drilling site. When the fracking process occurs, many chemicals can end up in water supplies. There have even been videos released of people setting their tap water on fire because of the huge amount of methane that has ended up in it. While methane in water supply has not been shown to have any impact on human health, the chemicals used for the fracking process can. Fracking has also been linked to the cause of earthquakes.[23] This is why small earthquakes in what should be geographically stable areas of the US such as Ohio, Pennsylvania, and Oklahoma can occasionally be experienced and are becoming more common in proportion to fracking expansion. As wastewater is injected below the surface for fracking, the water can cause geological faults beneath our feet to slip, creating an earthquake. Most of the time these are small, but there have been numerous reports of these actually cause damage to property. Additionally, they seem to be growing in size and frequency as the number of fracking sites increases. (See chapter 5 for similar impacts being felt by geothermal energy.)

Aside from the debate of the health impacts of fracking, it does play a significant role in climate change. Fracking targets natural gas, which is comprised of methane. Methane is 28-36 times more potent of a GHG than CO_2, so it takes much less of it to have a large impact.[24] Because of this potency, various studies have estimated that methane could be responsible for a significant amount of global warming (the EPA estimates 10% of US GHG emissions come from methane). As long as natural gas resources are handled properly, this isn't quite as dire of a problem as it is made out to be because when the methane is burned for heating and energy production, it turns into water and CO_2, vastly reducing the warming impact instead of just releasing methane. However, there are likely vast amounts of methane leaking from natural gas pipelines and drilling sites, though the data is not quite settled. Because many of the early studies on this topic were funded by the fossil fuel interests that own the pipelines under study, they are likely inaccurate or incomplete (in many cases, the company got to choose which pipelines were inspected under the studies). Various studies that have come out over the past several years have found methane leaks to be far greater than was originally projected. If it is true that methane leaks are extensive, then that could have major implications for climate change. On the other hand, natural gas has helped the US and other countries reduce their GHG emissions in recent history. This is why despite using more energy, US GHG emissions peaked back in 2007.

Coal obviously has its own dire environmental worries. While not causing earthquakes, coal mining is notoriously destructive to the environment and also releases some methane. (Coal mines in Appalachia, for example, are notorious for completely removing entire mountaintops, aptly called "mountaintop removal.") To be fair, pretty much any type of mining is destructive to the environment. Uranium, coal, and heavy metal mining can be some of the most damaging. In the case of uranium, this is of course mitigated by the fact that much less uranium is required than coal, but the radionuclides emitted also pose a threat not quite as prevalent in coal mines. In the case of rare-earth and heavy metals, recycling can drastically reduce the quantity of materials that needs mined, but you can't recycle coal.

Lifetime Economics*

The main reason we still primarily use fossil fuels for producing energy is simply because they're generally cheaper than other types of power sources. Compared to other consistent forms of energy (such as nuclear and hydropower), there are far fewer safety systems that need implemented and they can be built very fast at a much lower cost. The fuel is also much cheaper and easier to obtain than the enriched uranium for a nuclear reactor. Compared to renewables like solar and wind power, fossil fuels provide a much more consistent provider of energy, which will always be needed (at least until battery technology improves greatly). For these reasons, the more environmentally friendly alternatives to providing power have been highly neglected until recently.

However, fossil fuels are not cheap in the long run. Over time, fossil fuel plants (particularly coal) become less and less profitable as opposed to other sources over the same time period. Renewables have zero cost of fuel since it's provided naturally and although there is still fuel consumed, since nuclear demands far less and produces more energy, the costs are greatly offset. Therefore, despite spending billions on nuclear plants in the past, France has some of the lowest energy costs in the world and is the world's largest exporter of electricity, collecting nearly $3.5 billion every year just off electricity exportation.[25] (Not to mention having a very low carbon footprint for a country of its size, emitting less CO_2 than South Africa, Turkey, Poland, and Australia, among others. Despite having a substantially larger population, France produces less than half the CO_2 emissions of Canada.) This is contrasted to countries who rely heavily on coal, whose energy prices tend to run very high.

On top of the need for massive amounts of fuel, carbon pricing (often through the form of a carbon tax) policies are becoming increasingly popular around the world. These are meant to hinder processes that create volatile emissions and give non-polluting sources an economic advantage. Some have proposed taking it a step further and forcing fossil fuel operations to pay for the health damages they may indirectly cause on the general population through air-pollution, touted as realizing the "true cost" of fossil fuels. The combination of all of these policies, which would show the true cost of fossil fuels, would severely hinder their economic feasibility.

Additionally, many falsely argue that the only reason renewables are becoming affordable is due to government subsidies. When con-

---

* The economics of fossil fuels have been analyzed throughout this book in comparison to other energy solutions so this chapter will not dwell on it extensively.

sidering what fossil fuel corporations would be paying if environmental and human health damages were incorporated (therefore removing the indirect subsidies fossil fuels receive by not having to pay for damages they cause), the International Monetary Fund (IMF) calculated that fossil fuel subsidies reached $5.2 *trillion* in 2017—6.3% of global GDP—most of which went to coal and petroleum.[26] Of course, while fossil fuels receive indirect subsidies like this, renewables receive a plethora of direct subsidies. Considering US direct energy-related subsidies in fiscal year 2016, the EIA estimates that renewables consumed 45% of subsidies ($6.68 billion), coal 8%, natural gas and petroleum liquids 5%, and nuclear just 2%.[27] Additionally, smart grid and transmission accounted for just 1%, conservation 7%, and end use (things such as heating and lighting) was 42%.[28] This level of renewable energy subsidization then fell dramatically during the Trump administration and is likely to return once again under President Biden. Ironically, those who tend to be against the subsidizing of renewables tend to also be in favor of nuclear power, which is by far the biggest example of government subsidizing in energy history—if we exclude the true cost of fossil fuels.

Eventually, the world will run out of fossil fuels. It may be 30 years or 300 years from now, but eventually it will happen. There is a possibility, albeit an unlikely possibility, that if the world is continually dependent of fossil fuels and fails to find suitable solutions for producing energy, regional and global conflicts could erupt over resource conflicts.* Long before this point, however, fossil fuels will become simply uneconomical.

Global Usage Trends

In order to combat climate change, fossil fuel usage and production needs to be on the decline, starting yesterday. Unfortunately, the evidence is not clear whether this transition will occur hastily enough to avoid severe climate change impacts. While the Covid-19 pandemic

---

* This is backstory of the popular video game series, Fallout, where a conflict between the US and China over the world's last remaining resources—notably oil—eventually causes a global nuclear war.

severely reduced global fossil fuel production (especially oil), a rebound is already underway.

Coal. Due to the Covid-19 pandemic, coal production fell by about 8% in 2020—representing the largest drop since WWII.[29] In the US, coal fired electricity generation made up—for the first time this century—less than 20% of the power mix.[30] Expected coal demand reductions for the whole year was 25% in the US and 20% in the EU.[31] These trends were unique to the situation, and coal production will likely rise again after the pandemic.

Prior to the pandemic, the outlook for coal sent mixed messaging. While coal has been on the decline in the US and Europe, the increase in demand in other parts of the world (most notably in China) has been increasing at a faster rate. Coal power generation reached a record high in 2018, but then declined by 3% in 2019.[32] Additionally, in 2019 spending on coal power dropped by 6% and investment reached its lowest point since 1980.[33] Before this decline, in 2018, global coal capacity reached the peak of a long growth reaching 2,078 GW.[34] In the SDS, this amount should decrease to 1,636 MW by 2030, which is technically possible but highly unlikely.[35] While 2019 and 2020 were bad for the coal industry, global demand is not expected to be severely damaged. In 2040, coal demand is expected to be roughly the same as in 2018, with a negligible decline.* To comply with the SDS, coal demand needs to fall to just 38.5% of the 2018 demand by 2040.[36]

The most promising trend showing for the decline of coal seems to be the reduction of investments. As the world continues to tackle global warming, financial interests (such as banks, utilities, and private investors) are investing less and less in fossil fuels and more in renewable, carbon capture, nuclear, or breakthrough technologies. Coal is a prime example of this shift. Between 2012 and 2018, global coal supply investment was cut in half to just $80 billion.[37] Furthermore, the coal that is being invested into is generally for more efficient types of plants.

---

* The reason it does not increase significantly is due to several factors. Primarily, this is due to the onset of natural gas power, as well as the decline in price of renewable power, both of which will outcompete coal in the future. Energy efficiency improvements and global carbon policies are also responsible.

Natural Gas. Natural gas is controversial, to say the least. On one hand, it is a fossil fuel and releases GHGs (primarily CO_2 and CH_4). On the other hand, it is potentially the cleanest fossil fuel, is widely available to some countries, and efficient. Natural gas emits 50-60% less CO_2 than coal power plants, and 15-20% fewer GHGs than gasoline in vehicles.[38] * Because of this, there are those that see natural gas as a transitional fuel. It is to be used for the next several decades as a replacement for coal until renewables or nuclear can cheaply and consistently provide a baseload power supply. This reasoning is supported by a 2017 DOE report which found that natural gas, and not solar or wind, is already "The biggest contributor to coal and nuclear plant retirements." [39]

As discussed earlier this book, natural gas is often used for peaker plants because of how quickly its output can be ramped up or down. Because of the increasing expansion of wind and solar, natural gas use can be expected to grow alongside these technologies going into the near future. While it may not be desired to use natural gas for several more decades, the reality is that it is necessary. While it is a nice idea to instantly replace all fossil fuels and nuclear with wind and solar right now, the reality is that we need the consistency until other problems are solved. Additionally, many who pose the idea of not utilizing natural gas at all do not have effective alternative plans other than backup storage for renewables, which, as discussed in chapter 1, is not yet ready for wide-scale implementation.

Because of the numerous advantages that natural gas offers, it has been growing at a fast rate over the past decade (especially in the US) and is well positioned to continue doing so for an extended period of time. In 2018, natural gas consumption increased by an impressive 4.6% and was responsible for nearly half of the global energy demand increase.[40] By 2019, global natural gas production had increased by over 63% since the turn of the century.[41] Because of natural gas's conflicting qualities, it has a unique role to play under the SDS. Most technologies under the SDS either increase (such as renewables) or decrease (such as coal) their outputs over the goals time period. Natural gas has a different role to play. In the SDS, natural gas production trends upward,

---

* These figures do not take into consideration possible methane leakage, which, as discussed previously, could be significant.

reaching a peak in the late 2020s, then declines back down to slightly below its present position by 2040. This makes some sense because this should, in theory, give more time for energy storage development and the advancement of gen. IV nuclear power. Like many of the outlooks covered in the SDS, this scenario seems unlikely. In reality, natural gas production is expected to increase through the 2030s, to become over 37% greater than it was in 2018 by 2040.[42]

There is the potential that serious action to address climate change coupled with the (likely) discovery that natural gas leaks far more methane than is currently acknowledged could severely damage the fuels future outlook. Even aside from CH_4 leaks, natural gas fracking is extremely environmentally destructive, and many indigenous communities have been successful in recent history at preventing the drilling for this substance and others on their land because of the negative health effects and ecological damage caused.

Oil. Global oil production in 2018 was an impressive 98.3 million barrels per day (mb/d).[43] Over the whole year, that's more than 4.5 barrels per person! Oil did not escape Covid-19 though. Relative to 2019, oil demand was down 29 mb/d in April, and 26 mb/d in May of 2020.[44] Despite the setbacks felt across the industry by the Covid-19 pandemic and the need to reduce oil consumption (and therefore production) at a high rate due to climate change, oil production is expected to continue growing modestly for years to come.[45] While many developed nations' oil demand is plateauing (or in some cases slightly decreasing due to energy efficiency), global demand overall will increase due to developing nations. The only circumstance where oil demand falls drastically over the next several decades would be due to an unprecedented adoption of electric vehicles, hydrogen power or advanced biofuels, increased usage of carbon-free power generation, and a massive reduction in the usage of products that use petrochemicals (such as asphalt and plastics).

Summary on Fossil Fuels

There are several predicaments that the world finds itself in relating to fossil fuels that are each massively difficult.

While many developed nations have the ability to begin a moderately paced transfer away from fossil fuels, developing nations are not be as lucky. Historically, fossil fuels have been the main factor in driving the world towards economic growth and societal development, and continue to do so today. Because of this fact, there is a strong argument to be made that developing nations should be permitted to use fossil fuels as they develop. This can be mitigated by several factors, some of which might allow developing nations to leapfrog over fossil fuels to no-carbon technologies. Developed countries could take this opportunity to aid developing countries by providing low-interest loans, services, and policies that specifically target the addition of no-carbon solutions. This is particularly enticing because it gives an economic benefit to every party involved. The seller (in this case, a developed country) profits and creates jobs, while also enabling the buyer (a developing country) to not only mitigate issues such as air pollution, but will also simultaneously expand their entire economy. This allows for a positive feedback-loop of economic and political cooperation that benefits all parties involved and the environment. Technological solutions such as carbon capture could also permit for the burning of fossil fuels while not emitting any GHGs, which could greatly aid developing countries. China, for example, has already begun to invest large amounts of capital into developing African countries infrastructure and energy sectors.

Although today's discussion is about how quickly the world can transition away from fossil fuels, the reality is that energy transitions like this have historically taken place over decades or even centuries. In order to meet emissions goals and avoid the worst aspects of climate change, the world will have to transition at an unprecedented pace.

Fossil fuels are also an established entity, and the global economy is built around and dependent on them. Some of the biggest companies in the world (ExxonMobil, BP, Shell, Chevron)* are fossil fuel interests. Between 2007-2012, employment in the US oil and natural gas industry increased by 40%.[46] Just inside of the US electric power production industry, there were over 210,000 jobs that were directly related to coal, natural gas, or oil in 2018.[47] However, there were over 350,000 jobs in the same sector just between wind and solar (and an additional 92,000 who spent less than 50% of their time working on solar), with another

66,000 from hydro, 63,000 in nuclear, and 8,526 from geothermal.[48] An additional 240,000 work with natural gas, 184,000 with petroleum, and 36,000 with coal on transmission, distribution and storage of fuels. (The employment figures for renewables are negligible in transmission and distribution except for battery and pumped hydro storage—which employed and additional 70,00, but obviously this applies to forms of energy other than just with renewables).[49] Despite renewables making up *far less* of US electricity generation than fossil fuels, they already employ a large number of workers in the electric power industry. A "green revolution" is truly a job creator. Keep in mind this does not incorporate how many are employed by traditional fuels on the entire energy supply (many work with oil for cars, for example), but the comparisons are still clear. This global shift from traditional energies to non-carbon sources (particularly renewables in this case through the "electrification of everything") will create—and already is creating—a massive boost in employment.

Worldwide, millions are employed by fossil fuel industries. A key part of the energy transition will need to be to retrain the persons currently working in fossil fuel industries so their skills can be used for the clean energy shift, also while protecting these employees pensions as they leave the industry. For example, the technology between geothermal power and oil drilling is similar. There will be an increasing amount of people (assuming a decline in fossil fuel production) who work drilling for oil or natural gas who will be out of work. These workers can be quickly retrained for applications in geothermal power where they will not only have a steady, well-paying job, but the knowledge they have acquired over years of working for oil companies is not lost.

The transition of jobs from a high-carbon to a low-carbon economy can—at least in many cases—improve upon the low-carbon solutions while also providing greatly increased employment and a healthier environment and populous.

---

* Despite the massive impact of the Covid-19 pandemic on fossil fuel corporations, at the time of this writing, these four companies have a combined market cap of over $724 billion on the New York Stock Exchange.

Chapter 19

The World We Pass On

Climate Change, No Longer a Distant Threat

In 2016 Leonardo DiCaprio addressed world leaders at the UN in New York. His message? "The world is now watching. You will either be lauded by future generations or vilified by them. …You are the last best hope of Earth. We ask you to protect it, or we, and all living things we cherish, are history."

It's sad to say, but at this point in the climate crisis, we will not avoid some of the more drastic impacts of climate change. A widely accepted "tipping point" for global warming is at 1.5-2°C of warming (referring to pre-industrial temperatures as a baseline, which is most commonly referenced as between 1850-1900), which is the specific temperature that The Paris Agreement aims to stay below. We have already reached the point of 1.1°C warming, and the rate of warming is accelerating. We really do not have much time left to avoid 1.5 °C. The *Paris Agreement* was an agreement made within the United Nations in 2016 to mitigate the effects of climate change by keeping global temperature rise down, but its impacts have not been substantial since it does not legally bind countries to reducing their emissions.[1] Rather, it's basically on the honor system, which doesn't hold up in the age of fossil fuel dirty money (see chapters 6 and 17).

At the point of 1.5-2°C, most or all of the world's glaciers and ice caps will disappear, natural phenomenon (like wildfires) are projected to get much more frequent, and violent heat waves (with the addition of drought) are to become increasingly more severe, among the many other impacts that will be felt, as we will see later.[2] While global averages remain below this number, we have already passed this point on large amounts of the planet. Extreme areas of the planet have already reached increases of more than 2.5°C, with the fastest warming regions being the Arctics. Figure 19.1 comes from the Intergovernmental Panel on Climate Change (IPCC) Special Report on Global Warming of

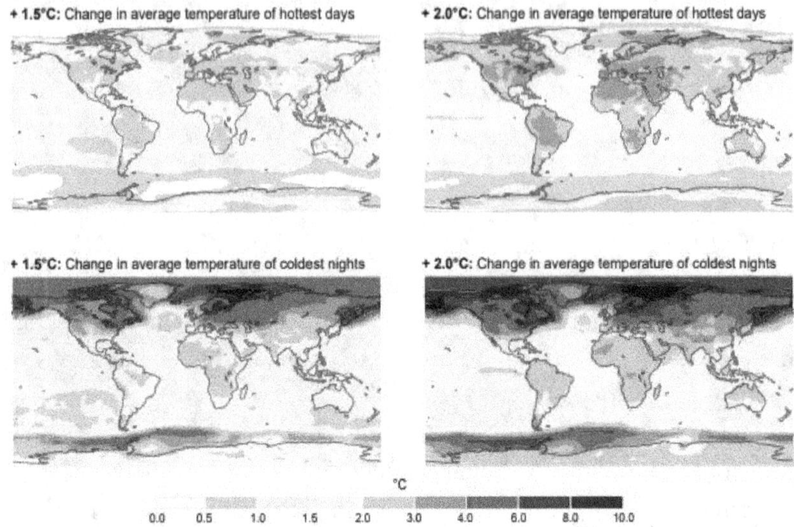

Figure 19.1: Impact of 1.5°C and 2.0°C global warming

Source: "Impact of 1 °C and 2 °C Global Warming.", IPCC, United Nations, 8 Oct. 2018, www.ipctc.ch/sr15/graphics/. FAQ 3.1, Figure 1. See citation "3" this chapter for more details. Taken with permission by the IPCC.

1.5°C and depicts how much average temperatures could be expected to rise around the world from the baseline previously discussed.[3] As pointed out, some areas could warm by up to 10°C.

And as the IPCC highlights, the best way to prevent this course is through the mitigation of the release of greenhouse gases into the atmosphere.[4] However, even if we completely stopped emitting GHGs right now, temperatures will continue to climb for decades. This is simply because GHGs take time to react with the environment and dissipate. In some cases, CO_2 can hang out in our atmosphere for up to a couple hundred years until the planet can absorb it completely, and some potent "F-gases" (Fluorinated gases such as HFCs, CFCs, SF6, & NF3, which can have warming potentials thousands of times greater than CO_2) can stay for thousands of years.[5]

It's not realistic to talk about how much the planet has already warmed in some cases, because our past actions will still affect us into the extended future. This is a large reason why the rapid development and implementation of CCS technology will be so important for mitigating warming. No matter how clean our energy gets, we will still have to face a substantial amount of climate change impacts as a result of the GHGs emitted from the past over the course of upcoming decades and possibly centuries.

As a human species, we will undoubtedly survive this continual heating, but it will inhibit our development in the future; primarily from things like food shortages, the spreading of disease and weather changes—including more violent tropical storms, forest fires, and extended droughts. This is still talking about if we stopped sending greenhouse gases into the atmosphere right now. And that is not going to happen. In fact, the year with the most GHG emissions worldwide was in 2019.[6] This was slowed by the Covid-19 pandemic in 2020 and 2021, but these emissions will rebound as economies ramp back up.

According to NASA projections, at the rate of change that we are transitioning towards a low-carbon economy, global temperatures could conceivably rise by a staggering 6°C by the turn of the century.[7] To put this into perspective, the global average temperature of the planet during the Last Glacial Maximum—when the ice sheets were at their greatest extent during the most recent major Ice Age—was approximately 6°C colder than the average temperature of the 20th century.[8,9] The cooling of the planet was so intense that it resulted in "vast glaciers, in places as much as several thousand feet thick, spread across northern North America and Eurasia. So extensive were these glaciers that almost a third of the present land surface of the Earth was intermittently covered by ice. Even today remnants of the great glaciers cover almost a tenth of the land." [10] However, unlike our current situation, this climate change occurred over the course of thousands of years, not a couple hundred. This gave life on the planet time to evolve and adapt to a new climate. We do not have this luxury.

As humans are impacting the climate more and more, we are not only increasing the temperature, but doing it at a pace far too fast for much life on the plant to adapt. A rise in temperature as extensive as predicted by NASA would almost certainly kill many things living on the planet, and the loss of crop yields in particular could have an immense impact on humans. Even the impacts that are being seen under our current warming (about 1°C) are having profound ecological implications. As stated, even if we completely stop using fossil fuels right now, we are still, quite literally, in deep waters.

Global Warming is Accelerating

The air we breathe is comprised of a hodgepodge of molecules, and we are able to measure exactly what the concentration of select molecules are. In this sense, we can see the actual increase in the amount of C2 and various other emissions from fossil fuels in the air as the composition of the air changes. Parts per million (ppm) is a measurement unit used to describe the consistency of something in a sample (typically a sample of a liquid or gas). Simply put: ppm describes how many molecules in a sample out of a million are comprised of a particular substance. For example, the first measurement of CO_2 ppm in the air ever accurately taken was in 1958, and it concluded that there were 313 molecules of CO_2 in a sample of 1 million air molecules.[11] Nowadays, that number has risen to almost 420 ppm due to the emissions of CO_2

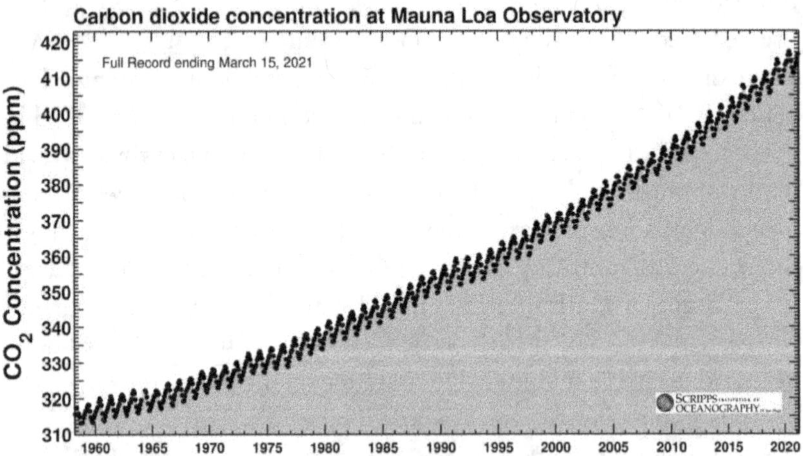

Figure 19.2: Carbon Dioxide concentration at Mauna Loa Observatory
Accessed on March 15, 2021. This resource is updated constantly. The Keeling curve trend is easily identifiable here. Source: "A Daily Record of Atmospheric Carbon Dioxide from Scripps Institution of Oceanography at UC San Diego." *The Keeling Curve*, U.S. Department of Energy, Scripps Institution of Oceanography, Jan. 2020, scripps.ucsd.edu/programs/keelingcurve/.

by humans,[12] identifying a 34% increase in the average amount of CO_2 in our atmosphere, meaning that CO_2 is just 0.04% of our atmosphere. When in terms of "Parts Per Million" these seem like insignificant numbers, but it is important to keep in mind that everything is relative. While water vapor is a GHG and often is far more common than CO_2 concentrations, water vapor is not the cause of the warming, but rather an amplifier.[13] Additionally, water vapor also has a cooling effect—largely from clouds reflecting incoming light back out into space, but even these clouds can have both a cooling and warming effect. Therefore, even though additional water vapor has a warming effect, it is a relatively un-important metric to watch because it is driven by other greenhouse gases. And most gases in our atmosphere have no warming effect. 99% of the atmosphere is either nitrogen or oxygen, both of which create no warming. An additional ~0.93% is argon, and helium which are inert and also have no warming impacts. Overall, assuming water vapor at 0% (because it ranges from 0-4 at a given time) the gas with the largest concentration left (by far) is CO_2. Without this 0.04% of our atmosphere, the planet would be freezing cold.

Another important factor to consider when dealing with the amount of CO_2 in the atmosphere is the Keeling Curve. Essentially, this trend demonstrates that during the winter, the amount of CO_2 in the atmosphere is higher than in the summer. This is because during the winter there are not as many plants alive to take in CO_2, and as the winter continues, plants die and release the CO_2 they had previously absorbed. In the spring and summer, plants grow back and begin to absorb carbon again. (CO_2 levels typically peak around May.) By measuring this trend that goes up and down throughout the seasons, we see a clear repeating parabolic pattern emerge in the aspect of ppm, rather than a straight or simple exponential line. Figure 19.2 shows all of recorded ppm measurements (as of March 15, 2021) by the monitoring facility on Mauna Loa, Hawaii, and although it goes up and down, it is easily identifiable that the average concentration of CO_2 is increasing.[14] (It is also important to remember that CO_2 concentrations depend on your location on the planet. A city, for example, will have a much higher concentration than the middle of the ocean. This is why the location of Mauna Loa is so useful, as its remoteness and elevation mitigates the data being skewed by local human populations and activities.)

Despite these worries, many governments haven't even stopped building new coal power plants yet. Not just in specific countries, but internationally, fossil fuels are still hanging around. Additionally, most power plants that utilize coal plants aren't even being shut down early but are being allowed to live their full life cycles and, in some cases, have even gotten extended life cycles. Coal plants often have life cycles of 40 years, which means that since we continue building them due to our lack of initiative to close them early, in a best-case scenario the world could hope to completely phase out coal over the next 40-50 years, and that's only if we stop building new plants now. At the United Nations Climate Summit in 2019, "65 countries and major sub-national economies such as California committed to cut greenhouse gas emissions to net zero by 2050, while 70 countries announced they will either boost their national action plans by 2020 or have started the process of doing so," [15] but at the time of this writing there are nearly 600 coal-fired units announced, or already under construction globally.[16] * There are no coal plants being constructed in the US, but this is because of the influx of natural gas production and use in power generation and heating. Globally, there are currently 6,600 coal-fired units in operation.[17] The vast majority of these plants are in Asia, in particularly China and India. In order to avoid disaster, China and India will need to further advance their renewable energy programs (though China already has one of the most ambitious goals on the planet and India is working on expanding its solar power though its 'One Sun, One World, One Grid' initiative).

Since 1880, the ten warmest years on record have all occurred since 2005, and seven of these were since 2014.[18] 2020 was the second warmest year on record, only very slightly cooler than 2016. As the planet warms, various effects will cause the Earth to warm even faster. This is generally referred to as a Positive Feedback Loop (PFL). This might sound like a good thing, but PFLs are the harbingers of destruction. A PFL in the sense of global warming is initially caused by warming but then goes on to generates its own warming, which in turn causes more warming and so on and on.

---

* Like how nuclear plants use multiple reactors, coal plants often have multiple "units."

A simple way to think about this is a snowball rolling down a hill. The snowball at the top of the hill needs a "push" to get started rolling down the hill. Once it is rolling down the hill, it will pick up snow. As it picks up snow, it will become larger, allowing it will be able to pick up even more, thus allowing the snowball to rapidly grow massive as it descends down the hill. In the case of global warming, the size of the snowball is the temperature of the Earth, the "push" is the anthropogenic release of GHGs, and the way that the snowball rapidly grows is a PFL. Unfortunately, we have so far not done a fantastic job at stopping the snowball from rolling down the hill. Hypothetically, and given enough time, it could become so massive that we can't stop it—which is called a runaway greenhouse effect. The snowball may eventually turn into an avalanche.

The reduction of GHG emissions is important to prevent and mitigate these PFLs. There are various positive feedback loops—most of which relate to additional releases of GHGs—that could accelerate the warming or climate impacts of the planet that are worth discussing. It is also worth mentioning that there are also some negative feedback loops that will act to cool the planet (such as the weathering of rocks), but overall, the impacts will be a strong warming effect.

Melting Permafrost. There are huge amounts of CH_4 and CO_2 stored underneath the permafrost (ground that is frozen year-round) on Earth. As the planet warms, the permafrost melts—in many cases, for the first time in hundreds or even thousands of years—releasing these GHGs, which will then contribute to warming the planet further. This could continue until all permafrost is melted, at which point the feedback loop would stop, but the damage would be dealt. One study found that by 2100, thawing permafrost could be responsible for 1 Gt of methane and 37 Gt of CO_2, a massive sum.[19] It is also crucial to keep in mind that the warming potential of methane over 100 years is roughly 30 times that of CO_2. Additionally, "…by the middle to end of the century the permafrost-carbon feedback should be about equivalent to the second strongest anthropogenic source of greenhouse gases."[20] There is no realistic way of preventing these emissions from escaping other than preventing the melting taking place by reducing our carbon emissions now.

Melting Sea Ice. As Arctic and Antarctic sea ice melts, there are multiple concerns relating to PFLs. First, ice and snow have a very high albedo. While the open ocean only reflects ~6% of sunlight, sea ice can reflect 50%, and sea ice plus snow can reflect up to 90%.[21] Because of this, melted ice that became seawater absorbs more energy and therefore warms the planet faster than if the ice was still there (think about how blacktop gets hotter than the sidewalk on a summer day). The warming water also then accelerates the melting of sea ice it is in contact with, and the process continues until there is no ice left. This mostly relates to the amount of ice that is present year-round in the Arctic, but the impact is also seen in ice coverage that grows and shrinks with the seasons. As observed by NASA, the yearly Arctic sea-ice minimum is declining at a rate of 13.1% per decade,[22] and Antarctic sea ice is decreasing by an astronomical mass of 150 Gt/y.[23] In Greenland, a depressing 278 Gt/y are lost. It seems very likely that the Arctic could experience its first sea ice-free summer by 2050.[24] Greenland actually has been recording lower local sea levels in recent history because the gargantuan size of ice sheets actually attract water by gravity, so them melting off is *reducing the gravitational pull* of the country and lowering local sea levels.

Additionally, there are concerns about methane that is trapped at the bottom of arctic oceans under sea ice being released into the atmosphere by warming temperatures—a similar situation to the melting permafrost previously mentioned. The current emissions from this may already be as high as 17 million metric tons of methane each year.[25]

Water Vapor. As mentioned previously, water vapor is the most common GHG, but is not responsible for global warming. Because it quickly cycles through the atmosphere, the amount of water vapor available is directly impacted by the concentration of other GHGs (mainly CO_2) in the atmosphere that produce consistent warming. As more GHGs are emitted, more water vapor will also be in the air due to evaporation. This is also the basic cause of some of the drastic effects of climate change such as more frequent and more severe hurricanes, drought, and flooding.

Tipping Points. The planet is reaching various tipping points in climate change. Tipping points are often the changes that result in positive-feedback loops, and have their own implications that will cause major destruction to the Earth's biosphere. For example, the Amazon will soon be (or potentially already is) releasing more GHGs than it is absorbing. The ocean currents of the Atlantic Meridional Overturning Circulation (AMOC)—which is responsible for keeping northern Europe warm—could collapse as the Greenland ice sheet melts away, leaving the UK and other parts of Europe to freeze. Once arctic ice is completely gone during the summer, it becomes much more difficult for it to grow back in winter. These tipping points, and others, will then extensively impact our global climate.

Consequences of Climate Change

The threats of climate change vary greatly and are of serious concern. The debate if it is happening or not is no longer relevant because of overwhelming evidence. Instead, a new debate has been brought forward: Should we do anything about it? This pivots on another crucial question: Do the implications of mitigating climate change outweigh the impacts of allowing it to happen?

The last chapter made the argument that even in the absence of climate change, the world should transition off fossil fuels. However, this obviously does not address the whole picture. In order to stop emitting GHGs, every aspect of our society (not just transportation or power production) has to change. This includes everything from how your food is grown to how your toothbrush is made. Because of how drastically society will have to change, the question of "Should we do anything?" is a rather reasonable question implying caution to the biggest undertaking in human history. This book has up to this point focused on the side of solving the issue, so we will now look at what some of the consequences of climate change could be if we do nothing. These events will only get worse with time, even with mitigation. At this point, mitigation will simply reduce how much worse they get. A combination of mitigation, adaptation, and international collaboration will be imperative to overcoming these threats.

Economic Consequences. Possibly the most intriguing reason for mitigating climate change could be simple economics. Aside from the improving economics of renewables, and the possible near-future break-throughs in nuclear power, there will also be major economic con-sequences to not mitigating climate change that incentivize a transition away from fossil fuels.

In 2020, there were twenty-two weather and climate disasters, each exceeding $1 billion and adding to a combined cost of nearly $100 billion, in just the United States.[26] Between 1980-2020 there was an average of seven events to this scale per year, but between 2016-2020 the average was 16.2 per year.[27] NOAA has recorded 285 individual billion-dollar disasters in the US since 1980, and 119 (over 40% of the total) occurred just between 2010 and 2019.[28] The total cost of all 285 recorded events is nearly $1.9 trillion.[29] By 2100, climate change could cost the US 10.5% of its GDP, and "[reduce] world real GDP per capita by 7.22 percent." [30] Another assessment by the EPA found that, in the US, annual losses could total over $500 billion per year from just the examined sectors (namely health, labor, coastal property, agriculture, and energy) by the end of the century.[31]

Additionally, an article released by the World Economic Forum in early 2019 mentioned that "More than 90 US coastal cities are already experiencing chronic flooding – a number that is expected to double by 2030. Meanwhile, about three-quarters of all European cities will be affected by rising sea levels, especially in the Netherlands, Spain and Italy," and that "coastal living is becoming a liability: the costs of sea-level rise could rise to trillions of dollars a year in damages by 2100." [32] In an extreme circumstance, sea level rise could render some coastal cities simply unusable. In some parts of the world, residents who live along coastlines and other high-risk climate change regions are already finding it difficult to get reasonable insurance coverage.

In 2016 the *EPA* released estimates on the social cost of CO_2. It es-timated the cost to have been $36 per metric ton (in 2007 dollars) in 2015.[33] Global fossil fuel CO_2 emissions were 35.6 Gt in 2015,[34] meaning the social cost could be calculated to be over $1.6 trillion (in 2021 dollars after accounting for inflation). Economic impacts will also be drastically amplified by incalculable social consequences and geo-political stability. Money spent on adaptation can greatly offset these costs in the short term, but they inevitably only delay the problem. For example, Miami has been raising its roads due to rising seas flooding its streets. Despite this, the raised roads and water pumps (which has cost hundreds of millions of dollars) is only going to buy the city 50 years, assuming that sea level rise is not faster than anticipated.

As the United States' Fourth National Climate Assessment puts it, "With continued growth in emissions at historic rates, annual losses in some economic sectors are projected to reach hundreds of billions of dollars by the end of the century—more than the current gross domestic product (GDP) of many U.S. states."

Migration and Regional Conflicts. There will undoubtedly be severe geo-political consequences of climatic changes. The World Bank found in a report that by 2050 there could be over 143 million climate migrants.[35] * This is far more than the amount of refugees being seen from Syria, for example, which is already stressing European immigration policies and politics.[36] Additionally, "forecasts by the UN International Organization for Migration posit that there could be between 25 million to 1 billion environmental migrants by 2050, moving either within their countries or across borders, on a permanent or temporary basis, with 200 million being the most widely cited estimate." [37] In 2018, there were 17.2 million new displacements caused by disasters, 16.1 million of which were related to extreme weather.[38]

As conditions in unstable parts of the world worsen (most specifically in Sub-Saharan Africa and the Middle East), internal conflicts over resources and power will become increasingly common. This could have profound humanitarian and geo-political ramifications. In 2018, there were an estimated 10.8 million new displacements caused by conflict, nearly 7.5 million of which came from Sub-Saharan Africa.[39] Of the 7.5 million conflict displacements in Sub-Saharan Africa, almost 2.9 million came from just one country—Ethiopia—and another 1.9 million came from the DRC.[40] By the end of 2018, there was a total of 41.3 million people around the world who were internally displaced due to conflict.[41]

---

* Broken-down by region: 86 million from Sub-Saharan Africa, 40 million from South Asia, and 17 million from Latin America.

As Sub-Saharan Africa and Asian populations are expected to boom, this will only exacerbate the risks to populations. A shrinking of available resources coupled with a booming population is a combination that is certain to produce the deaths of millions if regional stability cannot be guaranteed. This should not be another excuse for military intervention in every case (though it will also be needed in some cases), but rather rapid economic stimulation and humanitarian aid may be able to stabilize industrializing economies in ways that military intervention never could.

National Security. Climate change will influence national security concerns in a variety of ways. As the documentary *The Age of Consequences* highlights so well, climate change is already posing a threat to national security through increased terrorism, piracy, and general instability. Most notably, this is from the effects of sea-level rise, reduced fish and food security, mass migration, and changes in water availability—which may in some cases devolve into "water weaponization" where militant groups can take control of a water supply and therefore a whole region, as was seen in Somalia in 2011 by Al-Shabaab.[42] This was serious enough of a threat that "Climate change, lack of food and continued conflict involving water weaponization took an enormous social toll. Limited access of humanitarian agencies exacerbated by al-Shabaab's actions led to more than a quarter million deaths and hundreds of thousands of displaced persons." [43]

If climate change becomes bad enough, wars between countries could become common as resources become limited, and in the interconnected world of today, this can quickly draw in other powers such as the UN, US, EU, China, and Russia. While climate change in many cases is not the specific reason for conflict (it is often coupled by political tension or poor economic development), it is widely recognized as a "threat multiplier" allowing the proliferation of conflicts to become more common or simply giving the conflicts a reason or way to start.[44]

The threat that climate change poses to infrastructure are also of severe national security concerns not just at home but also abroad. In FY 2020, the Department of Defense spent $67 million on updating military bases for climatic changes or repair them due to climate-related damages.[45] The US Secretary of Defense, Lloyd Austin, released a statement in early 2021 firmly stating, "There is little about what the Department [of Defense] does to defend the American people that is not affected by climate change. It is a national security issue, and we must treat it as such." [46]

Ocean Acidification and Warming. "Global Warming's Evil-Twin" is what ocean acidification is frequently referred to as. One of the more severe specific impacts of global warming, ocean acidification is rapidly changing the ecology and chemical composition of the oceans. As mentioned earlier, the oceans absorb gargantuan amounts of CO_2 from the atmosphere, and globally, since the beginning of the industrial era, the ocean has absorbed over 525 billion tons of CO_2 from the atmosphere, currently around 22 million tons per day.[47] They absorb about 30% of CO_2 emissions,[48] making them the world's single largest carbon sink.

So how does this relate to acidification? Simply put, when CO_2 mixes with water, it lowers the pH, making it more acidic.* Since the Industrial Revolution, the average pH of the world's oceans has decreased by 0.11.[49] Considering the pH scale ranges from 0-14 (0 being the most acidic and 14 being the most basic, while 7 is neutral) this might not sound like much. The important distinction to be made is that the scale is logarithmic—like the Richter scale used for earthquakes. Otherwise, a 1 is ten times more acidic than a 2. So following the math, the oceans have become approximately 30% more acidic since the Industrial Revolution.[50] For reference, a drop of 0.1 pH of someone's blood can cause seizures, comas, or other major health consequences.[51] Furthermore, "Ocean acidification is occurring at a rate 30 to 100 times faster than at any time during the last several million years driven by the rapid growth rate [of] atmospheric CO_2 [which] is almost unprecedented over geologic history." [52]

As you might expect, this is having profound impacts on ocean life. This includes the massive die-off of coral reefs, crustaceans (acidic waters make it more difficult for these species to form shells), and some

---

* As you may recall from the introduction, CO2+H2O=H2CO3—more commonly known as carbonic acid.

fish species. While some species may be able to quickly adapt or actually benefit from this change, the net effect will be drastically negative.

It is also important to keep in mind that not only are the oceans acidifying, but they are also warming. This one-two-punch combination will have harsh impacts on marine organisms and human populations. These problems arise at a time when the world is also become increasingly dependent on seafood; global seafood demand was projected to be 154 million metric tons in 2011[53] and in the past 50 years, global seafood consumption has more than doubled.[54] With the exponential growth of our dependence on the market of seafood, the risk that is posed to marine life is extremely important, though it is highly overlooked. Many of the world's largest economies are very dependent on fish, particularly China, Japan, Indonesia, South Korea, and Norway. In 2011, China consumed 65 million metric tons of seafood, while the EU (the world's second largest consumer) only consumed 13 million metric tons.[55] Despite this, China only consumed 61.5% as much as South Korea and 83% as much as Japan on a per capita basis.[56] A major disruption to seafood markets could have obvious global implications, but clearly Southeast Asia will be one of the hardest hit regions.

It is also important to keep in mind that while the oceans acidify and warm, they are not changing at a uniform rate. Because of various factors, some parts are changing more rapidly than others. For example, the southeast coast of South America, parts of the Indian Ocean, the South-China sea, and both the east and west coasts of Australia are among some of the most rapidly warming regions. While the oceans are acidifying at different rates as well, they are acidifying relatively more uniformly than the oceans are warming.

Also, the amount of oxygen in the planet's oceans may have decreased by over 2% over the past 60 years due to warming,[57] though the implications of this are not yet fully understood.

Wildfires. In many regions, wildfires are becoming more powerful and more common. California, for example, seems to be almost constantly on fire these days. In 2018 alone, California wildfires cost the state nearly $1 billion and killed 93 people.[58] At one point in 2020, five of the six largest wildfires in California history were all burning *at the same time*, destroying an area larger than Puerto Rico.[59] Smoke from the fires was so intense it was noticeable on the east coast, where the sun was partially shrouded in some areas for several days (this then occurred again in 2021).[60] While things such as improved forest management could certainly help and need to be part of the solution, the main cause is clearly a warming planet. Blaming it on just "forest management" is nothing more than another red herring.

In Australia, the situation isn't looking any better. By February of 2020, a set of sweeping brushfires had burned an estimated 100,000 sq. km[61] (roughly 37,000 sq. miles), an area larger than the state of Indiana. Large parts of the Amazon Rainforest, Southeast Asia, and Africa are also burning. Unlike in Australia and California however, these fires are predominantly intentionally caused by humans. The "Slash and Burn" tactic has been extensively used in these regions to clear out forests for crops and animal grazing. While not the best for the environment, and may even be having impacts on global weather patterns,[62] this does reasonably benefit the local and regional economies. This is all to say nothing of the fact that the 2021 Siberian wildfires in Russia are—at the time of this writing—larger than all other wildfires on the planet combined.

The health impacts of wildfires through air pollution are also worth mentioning, and according to the EPA, "The effects of smoke from wildfires can range from eye and respiratory tract irritation to more serious disorders, including reduced lung function, bronchitis, exacerbation of asthma and heart failure, and premature death." [63]

Severe Storms. As more water vapor is in the air, storms are able to grow more quickly and to larger maximums. While there is conflicting evidence over whether climate change will cause these storms to be more common, it will certainly cause storms that do form to intensify to levels they would not have otherwise.[64] The impact from the majority of small storms may be negligible, but the massive storms (Sandy, Katrina, Harvey) that have been seen recently will likely have higher wind speeds, develop faster (giving less time for response/evacuations), and will be significantly wetter (meaning they release more water) than we are used to. Global warming will make storms of this magnitude appear more commonly.

Sea Level Rise. As previously mentioned, sea level rise could have massive economic repercussions for the world economy. Since the largest cities are almost always built along shoreline (40% of the US population lives in coastal counties),[65] this threat is one of the most existential. A tactical retreat may be useful in some area, prioritizing new construction inland rather than expanding along the coast, but in many areas, this will be insufficient to save the city. Adaptation will be imperative to salvage cities and reduce the damages that sea level rise will impose. Sea walls, water pumps, and raising roads may be sufficient in many areas, but there will still be substantial loss of property and revenue. This is only accelerated by the fact that various cities around the globe are sinking simultaneously as sea levels rise—such as Indonesia's capital city, Jakarta[66]—compounding these risks.

While melting land ice is seemingly the biggest contributor to sea level rise, there are also other factors at play such as thermal expansion of sea water. As with most things relating to climate change, small numbers can make a big difference. A 1-2 foot rise in sea level, for example, should not be underestimated. Global sea levels have risen by

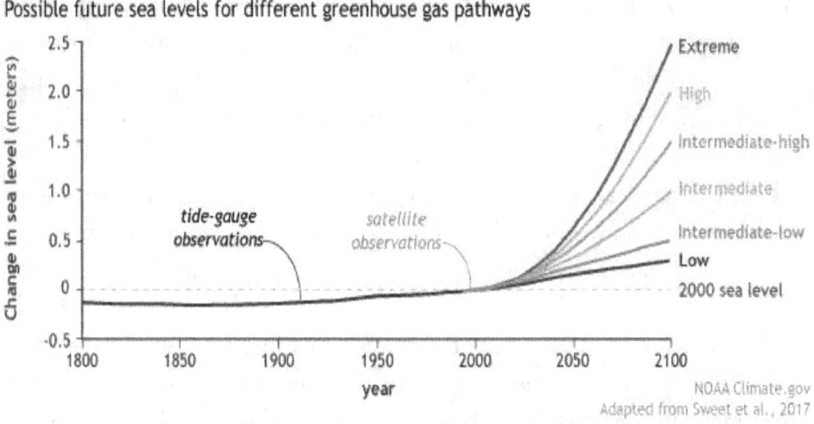

Figure 19.3: Possible future sea levels for different greenhouse gas pathways Source: Lindsey, Rebecca. "Climate Change: Global Sea Level." *National Oceanic and Atmospheric Administration (NOAA)*, 25 Jan. 2021, www.climate.gov/news-features/understanding-climate/climate-change-global-sea-level; see citation "67" from this chapter for more details. This chart was adapted for educational purposes, not for use for scientific reproduction.

8-9 inches since 1880 and will continue to rise this century, potentially (but unlikely) becoming up to 2.5 meters above 2000 levels by 2100 (see figure 19.3).[67]

Drought and Flooding. Droughts are likely to cause conflict and water shortages. While there will be more total precipitation in many regions, when the water comes it will be in sudden bursts instead of intermittently over the year. Because of the increased heat, water storages (such as lakes and rivers) are expected to dry up, and glacial and mountaintop ice and snow melt will decrease, severely impacting the water availability of many countries (including many in the developed world). When the rain is gone this will cause drought, and when the rain comes, massive flooding. The Southwest US has seen issues with drought in recent years,[68,69] and the severe drought in Syria may have helped spark its ongoing Civil War.[70]

Droughts in Afghanistan, coupled with terrorist organizations, are amplifying suffering in the country.[71] Afghanistan is experiencing an extreme drought (possibly the worst in over 900 years), and one of the only crops that can grow through its harsh drought is poppy, which is used to create heroin. Because of the drought, many farmers who were growing wheat or other food crops had to switch to poppy, which only requires one-sixth the water as wheat and is more profitable anyways. Part of the reason the US and NATO forces had such a hard time getting along with the citizens of Afghanistan was because of this issue. While western forces did not want to encourage or allow the growing of these crops, groups such as the Taliban would. This created trust between the Taliban and Afghans that was not possible with international forces and this trust clearly helped the Taliban take power in 2021. This obviously can then lead to more radicalism and terrorism. Because of climate change exacerbated drought (whether the drought was caused by climate change is impossible to tell, but we can tell if climate change made it worse), Afghanistan now produces around 90% of the world's heroin. Without climate exacerbated drought, heroin would be bread and the region would be more stable. Simultaneously, Afghanistan and other countries in the Middle East and Africa have seen massive flooding alongside these droughts, which has killed hundreds and displaced orders of magnitude more.

In 2010, a drought that was likely amplified by climate change in Russia caused the country to temporarily stop exporting wheat, which was a contributing factor to the Arab Spring.[72] (Middle Eastern countries, Egypt in particular, depended heavily on Russian wheat imports.) Drought is a prime example of how climate change acts as a threat multiplier.

Heat Waves. Heat waves could potentially become much more deadly due to the increased average temperature. This is only inflated when considering the overall increased humidity. A high humidity makes it difficult to cool off through sweating, so the combination of higher temperatures and humidity can make heat waves much more lethal in the future. This will obviously not be as much of a problem in developed countries with easy access to air conditioning and modern housing, but developing countries (especially those near the equator) could have significant problems. Extreme heat does not often grab headlines, but in reality, heat waves are responsible for more fatalities in the US each than all other natural disasters combined.

Species Distribution. Some projections by NASA[73] and more recent calculations by the UN[74] point out that by the end of the century temperature increases may be 5 °C or higher by the end of this century—up to three times greater than we had originally planned to avoid. Many counterarguments point towards the fact that it doesn't seem logical that a few degrees change in the average temperature would have such an impact, but the impact is scientifically and historically proven. In fact, in the Earth's past, "a one- to two-degree drop was all it took to plunge the Earth into the Little Ice Age." [75] The Little Ice Age occurred during human recorded history—so we can reasonably extrapolate what the effects of something similar may look like today—and had drastic effects, especially between the 17th and 19th centuries. It was reported to cause massive food shortages around the world, most notably in Europe where it was likely a contributing factor to the devastating Thirty Years' War, which killed millions. If the effects of a 2°C drop in temperature was enough to impact the planet this drastically, a 5-6°C increase from the same base point could be catastrophic. For a more relatable perspective, a rise in the human body's temperature from the normal 37°C (98.6°F) to 39°C (102.2°F) is obviously seriously harmful. An increase of 5°C would be equivalent to running a fever of 107.6°F. As was previously mentioned, a drop of 5°C in Earths past was enough to put us back in a major Ice Age.

As the planet warms, this dramatic shift is causing animals to occupy areas that they otherwise would not be in. Marine animals are going towards the poles or deeper into the ocean while land animals are going to cooler weather or higher in elevation. This shift in the regions that species occupy will have massive impacts on those who not only rely on these animals for food and work, but also rely on the role that these animals play in ecosystems.

Spread of Disease. There are a plethora of ways in which climate change may proliferate the spread of disease.[76] This is supported both by scientific research and throughout history. For example, a WHO report "…found a robust relationship between progressively stronger El Niño events and cholera prevalence in Bangladesh, spanning a 70-year period." [77] As is often common with disease, mosquitoes are one of the biggest worries. As the seasons get warmer, the insects are able to migrate farther into regions that were previously too cold.[78] Furthermore, the mass migration of both land and marine species may result in increased zoonotic disease transmission pathways, something that I think we can all agree is a bad thing now that we have lived through two years of this pandemic.[79] There is the possibility that diseases frozen away in permafrost that may be extinct otherwise could return and pose a new threat that our bodies are not equipped to handle, though this hypothesis is highly contested. (It would make for a good apocalypse movie though.)

Desertification. Desertification of land (particularly around the equator) is a serious threat to various regions, most notably North Africa, the Middle East, and China.[80] While increased desert land will provide a negative feedback loop (deserts have a higher albedo than plant coverage), it will also come with consequences such as increased food and water insecurity, dust storms (which are already having significant impacts on the air quality of Beijing), and overall forced migration.[81] And of course, deserts are not capable of storing carbon in their soil like foliage, which will contest the negative-feedback from the increased albedo.. The WHO lists the following health implications of desertification: higher threats of malnutrition from reduced food and water supplies, more water-and food-borne diseases that result from poor hygiene and a lack of clean water, respiratory diseases caused by atmospheric dust from wind erosion and other air pollutants, and the spread of infectious diseases as populations migrate.[82] China is already seeing massive issues with the expansion of the Gobi Desert, which, as reported by The New York Times, may be expanding by over 1,300 square miles every year.[83]

Food and Water Security. All of these combined factors will have major impacts on food security globally. Food and water shortages are large causes for social instability and radicalization. Therefore, food security poses one of the biggest risks to national security—for all countries—today. Even the Pentagon recognizes the risks of climate change.[84] One of the countries that could suffer the worst food security impacts is the US. The ongoing droughts in the Midwest of the US have severely hurt crop production and several years of drought in a row more East where most of the US's food is grown (such as in Iowa, where climate impacts like drought, flooding and heat waves are already being seen) could have major ramifications around the world. Short food and water supplies will (not could, *will*) result in social uproar, increased violence, civil wars, and if it is widespread enough, massive famines. Beef production will be one of the first and most impacted food industries as cattle require massive amounts of land, feed and water (a huge amount of the food grown in the US goes to feeding cows or chickens).

Ecology. Global ecology is already suffering the impacts of humans and climate change. In *The Sixth Extinction*, Elizabeth Kolbert travels around the world and showcases this to the reader with heartbreaking detail. Kolbert continually maintains the argument that humans have changed the world in such a drastic way that we have created for the planet an entirely new epoch, what would now be described as the *Anthropocene*.

Regardless of if we should consider today to be the Anthropocene, humans are without doubt driving the machinery of destruction for much of the planet's life, and many think that we are now in the midst of the sixth mass extinction in Earth's history. Overhunting, disturbance of ecosystems, industrialization, pollution, overpopulation, the extraction of more resources than the planet can handle, and climate change are all simultaneously damaging ecology, and we do not seem intent on stopping. Some of the most at-risk species are amphibians, corals, crustaceans, arctic animals, and various types of plants such as conifers.

Opinion time: I tend to hold an unpopular position on this issue. In truth, I tend to not be concerned about the polar bears or frogs. I am much more concerned about the impact on humans, and therefore am not as worried about ecology. However, it would be ignorant to ignore several factors. First, ecology plays a crucial role for humans, and without a proper ecological balance humans cannot continue to operate the way we do (many will argue that we can't continue this way of life either way). Second, humans get massive benefits from ecology. Most of our medicines come from plants (or chemicals originally discovered in them), and in order to continue advancing society, a healthy biosphere will be imperative for maintaining societal harmony. Lastly, one might even say that we have a moral argument to protect life on Earth. As far as we can tell, there is no other life in the universe. For at least as long as this stands true, it would be shameful for us to allow for life to perish from the universe. As the most dominant species on this planet (and possibly in existence), it would be shameful to destroy any significant portion of life on this planet.

So while I don't consider myself an environmentalist, I am a humanist and an ecomodernist. In order to protect humans, we also have to protect our natural environment. At the very least, there is a selfish reason to preserve our planet.

Summary: Is All of This Really Necessary?

Yes. To the vast majority of the world and the scientific community this is now obvious. This is why I have not been concerned with discussing the reality of climate change for most of this book. Most who still refuse to believe can likely simply not be convinced. Arguing about whether it is real or not is, at this point, frivolous. What is important now is that we have discussions about the best and most realistic approaches to reaching our goals.

Conclusions

Overcoming The Great Filter
A New Path

Where is everybody? These are the simple but famous words of Enrico Fermi. Fermi—whom you may recall from earlier this book—was a brilliant American-Italian physicist who gained popularity from playing a very important part in transferring the world into the nuclear age, but when Fermi asked, "Where is everybody?" he was talking about something very different from the nucleus of an atom. He was referring to the apparent lack of extraterrestrial life in the universe.

No matter your stance on extra-terrestrial life, it is hard to deny that life of some sort may exist in the universe other than on Earth (at least, from a non-religious perspective). Life on Earth may have been possible billions of years ago.[1] With the endless number of planets and the incredibly long life of the universe, it feels extremely likely life has and will form on other planets. But if millions (in all likelihood, billions) of planets host even basic forms of life, it is not then far-fetched to assume that intelligent life like us is likely to have existed sometime in the past or is existing in the present. Despite this, as far as we can tell, there isn't any. This is what has been coined "The Fermi Paradox." The Fermi Paradox represents the contradiction between the extremely high, statistical likelihood of extraterrestrial life and the apparent lack of it. In other words, if the universe should be teeming with life, "Where is everybody?"

A well accepted theoretical solution to this paradox is what is often called "The Great Filter." It makes its case as though there must be a point in life's development that is so hard to overcome, life almost always (or always) dies out. This would presumably happen before they are able to extend off their host planet and expand into the universe—the failure of which would probably lead to extinction through cosmic events such as asteroid impacts. Some examples of where this filter may lie could be the development of single-celled organisms, the development of intelligent life, language, the formation of society, or getting off the host planet. The problem is, we have no clue what or when this filter may be because we have nothing else to compare ourselves to.

As we draw closer to our own space age and reaching out onto other planetary systems the question remains: Are we in front of or behind this "Great Filter"? Have we already extended beyond the point where other life fails, or are we yet to come across this filter in the future? It is my personal belief that we are already currently inside this filter, and it's called climate change.

Let's make something clear: Man-made climate change is not an existential risk (at least, not for humans). An existential risk is technically a risk that could potentially lead to complete extinction (or at least, dwindle the population low enough so that it cannot reproduce effectively, creating an eventual but not immediate extinction). Even a nuclear war *might* not be an existential risk, as some blips of humans would be likely to survive in bunkers for extensive periods of time, before eventually coming outside and repopulating the planet. A true existential risk would be something on the scale of a black hole consuming the planet. The sun expanding over time will make Earth inhospitable. An infectious disease that goes undetected but sterilizes large quantities of the population could lead to a slow extinction of the species. In the far future, self-replicating nano-bots could get loose and consume the planet. Or possibly nuclear war will end up doing us in after all. So, while climate change is not existential, it is—what I consider— (potentially) *indirectly existential*. In this category I would also list nuclear war, supervolcanoes, and major asteroid impacts. To a species other than humans most of these truly are existential, but we are resilient, and I have faith that at least some of us would survive in bunkers or on another planet in a scenario like this.

This deserves some context. Modern human civilization is only possible because of climate change. Without climate change, we would still be living in caves and chasing mammoths. Once the planet warmed, we were able to grow crops more easily and more plentifully and expand our habitation. The reality is not that Earth has been kind to us; it is that over just a few thousand years humans have taken advantage of an inter-glacial period of a warm and uncommonly stable climate. If the Earth were cooling instead of warming—even if it were not man-caused—at a rapid rate that was difficult to adapt to, we would still try to prevent it. This is because we are living in a "sweet spot" right now. If the planet stayed cold—or became too hot—civilization would not have been possible. If civilization were not possible, we would never be able to make spaceships to colonize other planets. One of the popular theories for the Fermi Paradox is that any result that does not allow for a society to colonize other planets could likely be described as indirectly existential because this is (seemingly) the most important step we as a species have left to guarantee our indefinite existence. If you are unable to leave your host planet you will, eventually, face a direct existential risk. Inhabiting multiple planets does not eliminate this threat (there will always be concerns about things such as the ultimate end of the universe), but it drastically increases the chances for overall survival and species longevity. In the distant future, we will be able to terraform planets to make them more suitable for us to inhabit, which will again drastically decrease the threats of existential risks.

Humans are interesting creatures. We create solutions, and then those create other problems, as Elizabeth Kolbert depicts in her book *Under a White Sky*. What is almost always true, however, throughout human history, is that the problems our solutions create are not as bad as the original problem we started with. Take climate change as an example. You could argue that climate change is the worst ecological issue we have ever seen as a species, but on the flip side, the burning of fossil fuels has largely eradicated extreme poverty and led to unforeseen benefits to our species. I would say that the large removal of suffering from the lives of countless billions of people in the past and an infinite amount in the future is worth the toll of climate change, especially if we do indeed solve it. After climate change, we may realize that our renewable energy solutions created another ecological problem in our endeavors to mine rare earths from the ground. And so, the cycle continues.

Civilizations fall when they fail to rise to the challenge, whether that is because of political or social factors, inability to establish peace, or just plain ignorance. The single exception to this is, once again, climatic changes. As a threat multiplier, man-made climate change is yet again an indirect existential risk to individual civilizations. This can be seen frequently over history. Some prime examples are the Akkadian Empire and the Mayans, both of which likely collapsed at least in-part due to climatic changes.[2] If we fail to inflict major change over the next century, we could be risking our species survival for an extended period of time. In the most extreme cases, a runaway greenhouse effect could become directly existential, though this could take thousands of years.

For a real demonstration of how harshly CO_2 affects our atmosphere and global warming, we can look to another planet in our solar system. Venus was projected to once have been possibly habitable and have liquid water. Today, because of a runaway greenhouse effect, CO_2 makes up 96.5% of Venus's atmosphere and the planet reaches temperatures of up to 880 degrees Fahrenheit on the surface (hot enough to melt lead), with an atmospheric pressure being about 92 times that of Earth—approximately the same as being one mile under-water.[3] This concentration of CO_2 in the atmosphere is the reason why, despite being almost twice as far from the Sun, it still has surface temperatures higher than Mercury.[4] This is an extreme comparison, considering CO_2 only makes up about .04% of Earth's atmosphere,[5] but I think that it still draws a useful comparison. It is also important to keep in mind that while this is an extreme circumstance, it is physically possible (and some extremists might say plausible at our current rate of GHG emissions) for Earth to undergo a runaway greenhouse effect, as did Venus which once had liquid water on its surface and likely would have been welcoming to life as we know it.

The continuation of the use of fossil fuels and failure to implement clean energy on a global scale is among the most difficult problems humanity has ever seen. Not because of physical limitations, but because of our own psychology. The World Wars of last century showed humanity their own potential for devastation and pure cruelty, and the Cold War threatened a nuclear Armageddon. But these problems were in our face. They were things that we wanted to fix because we felt they threatened us directly. A nuclear bomb destroying a city is a very observable thing, and therefore we took great action to prevent that from happening. Climate change is not as easily observable, and, therefore, we have been doing much less to fix it than we should be. Unfortunately, humans just aren't good at thinking about problems in a long-term sense. We have the ability to feed the global population, for example, but it's not a problem that we see for ourselves and so most of us don't even think about it most days. Climate change is the same. In more ways than one, climate change is more of a psychological and sociological challenge than it is a technological one.

<u>In the face of all of this, we must create a new path forward:</u> This path must be fundamentally different than the ones we have chosen to seek in recent history and provide us with the ability to not only combat climate change but to also improve our way of life in fundamental ways. The purpose of this book has not been to lay out exactly what this path will be, as some may have expected from the introduction, but rather to show what some of the boundaries of operation within this path may entail. Various individuals and organizations have taken to attempting to lay out a plan and framework for this new path (such as Bill Gates in How to Avoid a Climate Disaster and Hal Harvey et. al in Designing Climate Solutions) , though none of them (to my knowledge) have formally recognized that they are working to create a distinct path that operates outside of the current two-path system the world currently debates.

Though the objective of this book has not been to bring attention to specific actions that this new path should operate under, there are three items that I believe to be safe to assume as the foundational bricks for our course of action. While the meat of these items as described shortly will undoubtedly be contested, I believe you will be hard pressed to find a professional of reliable position (at least one that is not a Libertarian or Anarchist) that disagrees with any of these focuses on a fundamental level.

1) Government R&D into technological solutions, as well as aid for implementation of solutions currently available. The main focus of this book has been to highlight issues, misconceptions, and challenges that face the global energy transition. Government funding will be imperative in order to solve these issues. This writing has highlighted a number of specific places where this funding should go, and we can summarize them again here.

Energy Storage technologies development will be imperative in order to rapidly expand the use of solar and wind on the scales which are required to avoid some of the worst impacts of climate change. Without an effective energy storage system, it is likely that these sources of energy will be expanded anyways but will come with various negative economic impacts as seen in California and Germany.

Energy storage could also help other clean technologies such as current nuclear technologies, though it is not commonly discussed. Modern nuclear plants are terribly slow at ramping up and down production of energy. This means that they end up running at a consistent power for extended periods of time. Many nuclear plants will run at a specified energy output (typically between 90-100% maximum power) for months or even a year at a time, with little to no fluctuations. This makes nuclear a horrible choice to have your grid depending solely on due to constant fluctuations in energy demand throughout the day. Because of this, nuclear plants are sometimes backed up by dirty peaker plants. An effective energy storage system could allow for an increased reliance of both nuclear and renewables while simultaneously reducing dependence on peakers. A few countries such as France do operate nuclear plants capable of load-following, but these are less efficient and generally not as effective as other load-following sources. (However, many gen. IV reactor designs appear to be much more flexible than existing nuclear plants.)

Some top candidates for additional funding and research for energy storage should be Li-Ion batteries, compressed air storage, and thermal batteries (particularly through the use of molten metals such as sodium, and molten salts.)

Energy Transmission will also become increasingly important for a grid that relies more heavily on wind and solar. An order of magnitude improvement in the viability of transmitting electricity could completely revolutionize the way we view how the power grid works. A smart grid in developed countries deserves massive attention, as it is all but guaranteed to more than pay for itself in a relatively short amount of time. In under-developed or developing countries, it may be better to focus on micro-grids (grids that are very small, powering an individual community or just a couple homes per grid) through the use of solar power especially, so that they can leapfrog over the massive capital required to create nationwide power grids.

Money invested into energy transmission infrastructure should focus especially on smart grids, as they may have the biggest "bang for their buck." High-voltage transmission lines will be important for efficiently transmitting energy without loss of energy, and China can already be seen investing heavily here. Other focuses could include underground transmission lines and micro-grids where applicable.

Fourth Generation Fission and Fusion Nuclear Technologies must be a major focus of support. Specifically, thorium fuel cycles and small modular reactors provide the opportunity to generate clean energy for millennia, while simultaneously improving national security and public health, reducing current inventories of nuclear waste, and expanding our options for power.

Specifically, Molten Salt reactors operating on a thorium fuel cycle, Fast-Breeder reactors, Pebble-Bed fueled reactors, Gas-Cooled reactors, and various nuclear fusion projects should be encouraged. While ITER is on track to fire up this decade, an increase of funding from all participants on DEMO (should it be deemed plausible) should be encouraged in order to speed up the process so that fusion can have an influence on the energy transition by mid-century. Personally, I believe that the amount countries (especially the US, as it has long been a leader in nuclear technology) spend on innovative breakthrough nuclear technologies should not be measured by the millions of dollars, but rather by the tens of billions, as the technology will more than pay for itself in the end.

Solar and Wind Power will be imperative for the energy transition of this century, especially from now until mid-century. These technologies have massive benefits that include increased employment in well-paying jobs, massive public health benefits, lower ecological destruction when compared to fossil fuels, and national security. To not incentivize the implementation and development of solar and wind at this point in the technologies' maturity is foolish at best and responsible for countless loss of life and economic disparity in actuality. The degree to which our energy supply should rely on these is certainly up for debate, but all people of reason should be able to recognize they have a significant part to play.

As the solar and wind industries have already taken off, government aid for R&D will have a limited impact when compared to R&D coming from the private sector. Something that governments can do is to prioritize their implementation over other energies and provide subsidies (to a limited extent), which will affect individuals looking into renewables. For example, a small government subsidy to encourage the use of rooftop solar could have net economic benefits if the energy would have otherwise come from a coal plant. In a case like this, the subsidy more than pays for itself in many circumstances.

It will be important to improve storage technology one-step *ahead* of solar and wind rollouts, as the inability to store the energy these generate could result in huge losses to the global economy.

Transportation Infrastructure will need to be of utmost importance in all countries. In the US and many European countries, infrastructure that was built immediately after World War II is today seriously degraded. A massive plan, such as many aspects of Joe Biden's proposed infrastructure plan, is exactly what is needed in order to repair out-of-shape infrastructure and "build-back-better." Many who study China's politics and economy understand that when Beijing's economy is stagnating, they invest heavily into infrastructure in order to kickstart the economy. This has worked incredibly well and helped the country to grow its economy at a rapid pace. Something similar may help jumpstart the stagnating economy of the United States by providing jobs and allowing for future savings via costs that will not be incurred. For example, a poor road will cause more damages to vehicles than a road that is well maintained. Repairing existing infrastructure will have massive cost savings for the future on the population as a whole, though the upfront costs are significant.

There is no shortage of ideas of ways that we can invest into infrastructure in order to provide for a green way of getting around, while also improving the health of the economy and creating a multitude of well-paying jobs. This includes a large incentive on electric vehicle charging station infrastructure, designing roads with features that will be useful for autonomous vehicles in the future, so that they will not need to be completely replaced far down the line, improvements in public transportation and its availability, high-speed rail projects (it is likely this is not going to be possible in the US, but many other countries have massive potential for HSR), and an improvement in both light and heavy rail so that more commerce can be transported by rail instead of trucks. This last point is important because not only is transport by rail less carbon intensive than by road, but this would save gargantuan amounts of taxpayer money via road damages that are avoided, which comes with additional savings of its own.

Other Focuses of Solutions should include government subsidies on electric vehicles and low-carbon alternative materials (while this is not very viable yet, it could have a huge impact in coming decades), a constantly increasing push on energy efficiency for appliances and housing, and a large push into investigating and developing a circular economy. The circular economy focuses on reuse of materials much more heavily than today. This will be important in the future as electronic waste, rare-earth minerals, and heavy metals become more difficult, expensive, and environmentally damaging to mine. Instead, it will soon be more practical to rely much more heavily on recycling these products so the materials can be used repeatedly for an indefinite amount of time. A circular economy will require a large push for development of innovative recycling practices, which we will need sooner than most think.

2) Economic regulations on carbon and other pollutants that are currently felt by society as a whole. As discussed earlier in this book, if the true costs of carbon and other pollutants emitted from fossil fuels were incorporated into their upfront costs, there would be no debate over how costly they are. An incorporating cost for these would massively incentivize a transition to cleaner fuels, though substantial political opposition is inevitable. I am no economist, so I won't linger on this topic for long, but I can confidently point out a couple specific steps that need taken.

A Carbon Tax might not be needed at this point in the energy transition but would undoubtedly speed up the process. The two primary functions of this that have been the subject of debate are Cap & Trade and Fee & Dividend. I will not concern you with the specific details of each, but what is important to understand is that a cap & trade system sets a cap on emissions and allows a company to sell or "trade" tickets which allow for a limited amount of pollution between companies while a fee & dividend inserts a tax somewhere on the supply chain of fossil fuels and distributes the collected tax revenue proportionally over the entire population. In any carbon tax system, the tax or fee should be included as close to the source of production of fuel as possible, rather than at the source of consumption. And that is basically just a fancy way of saying that the tax should be implemented at coal mines rather than coal power plants, or oil rigs and not the gas station, for example. This helps to prevent the formation of loopholes in the system and is overall more effective and less prone to corruption.

There have been problems with industry and lobbyists gaming cap & trade systems in the past to advantage themselves, and this is why many experts now back the aforementioned dividend system. Whatever happens, the implemented tax should not be stagnant. It should start at a modest price and gradually rise year by year on a cemented schedule dictated by law so that industry can predict future expenses and not lobby for advantageous changes when political power shifts. This will eventually force the costs of carbon to outweigh any economic incentives to using fossil fuels while also slowly shifting the industry so that it has time to either adapt or prevent the unemployment of an entire industry at a single moment. It is reasonable to expect that under such a system many fossil fuel companies will either rapidly transition into "energy companies" (as is already being seen with some big oil companies like ExxonMobil) or go bankrupt.

Realizing the True Costs of Pollutants on Society would bring about a swift decline in fossil fuel use. As discussed throughout this book, the social burden of harmful emissions from fossil fuels on society is horrendous. To consider air pollution a plague on society would be an understatement. Incorporating just the public health costs onto fossil fuels would make their price skyrocket far past what could be provided by any other form of energy, and this is to say nothing of the environmental costs. Admittedly, an economic tax that fully realizes this will probably never happen. However, it seems safe to assume that future generations' textbooks will one day recognize the atrocities committed via this form of energy. What can surely be done is an expansion of public understanding of this scourge to help break down social and political barriers to fossil fuels' phase-out from the global economy.

3) Improvements in education, including public understanding and the combating of disinformation. We might not be in our predicament if it were not for poor public education. Unfortunately, a few main factors relating to education have led us to this point. First, scientists tend to not be so great at communicating ideas to a broader audience. While there are some exceptions of this (most notably Carl Sagan, Neil deGrasse Tyson, and Bill Nye the science guy—who is an engineer by profession, but we will let that go)* the majority of scientists seemingly tend to prefer to stick to the science. This unfortunately leads us right to the second problem. Because scientists are often not interested in communicating their studies to a wider audience outside of the scientific community, this leaves mainstream media and politicians to explain it, which instantly makes it politicized. This politicization of science is, in my opinion, one of the most unfortunate developments to education since post World War II. What can we do to correct our course in education?

Getting Politics Out of Science Communication is the obvious first step here. The science of climate change has had large consensus inside of the scientific community for over thirty years. Despite this, it has not been taken seriously in public perception until very recently. Establishing independent and non-politically tied pathways for scientists to communicate to the general public easily and broadly should be of major effort. This transcends just climate and energy, but rather is important for a vast quantity of subjects currently wrapped up in political discourse today. Of course, scientists don't always agree, but allowing for easy communication should still get general thoughts across. Admittedly, the CDC was supposed to be a good example of this but

---

* I would also like to appreciate here the scientists who I interviewed for this book who were excellent at describing many topics to and overall very helpful in my research.

during the Covid-19 response they failed horribly at keeping politics out of the science. It is something easier said than done.

Improving K-12 Education and Scientific Literacy will be important as well. K-12 is a crucial point for getting youth interested in science and establishing a trusting relationship between scientific understanding and an individual's personal, political, and religious beliefs. Long time failure in K-12 to establish this trusting connection is a major contributor to the current situation. Encouraging and teaching an ability for students to be able to do their own research that comes to reasoned conclusions through exploring scientific sources outside of political influence could have a substantial lifetime impact on how a person researches and comes to conclusions in the real-world. This in theory would also help to diminish the effects of disinformation by a great deal. Additionally, all scientific research should be publicly available at no cost to anyone on the internet—something many scientists themselves argue for—and our school systems should ensure that the average person understands how to properly utilize this resource.

Substantial Investments into Women's Education in Developing Countries would have profound impacts for the climate and global society. Women in developing countries are often left to receive little or no education. The connection between women's education/financial stability and the number of children they have is well studied and understood. Fewer children at a higher income would not only help propel these countries out of poverty, but would massively alleviate potential stress on the planet's resources (and carbon budget) while simultaneously improving projecting women's rights in these countries in a positive direction.

To not change our ways is to destroy our human way of life. No matter how bad we hurt the environment, I believe people will continue to live on. We are too advanced to simply die off from a problem like this, but the world left behind could be a terrible dystopia. It would be an impoverished world. A planet overrun with natural disasters. Massive famine felt on a huge scale. Disease running rampant. Society torn apart. Conflicts raging across the globe in never-ending bloodshed over diminishing resources. The human race will continue on, but we might no longer advance, and what is our purpose if we are not advancing?

We would be so severely crippled by our environment that *thriving* would be much more difficult, nay, impossible. Disease, famine, drought, natural disasters, and wars will rip society apart. Economic and social stability will be unachievable due to unpredictable crop yields and global conflict. Humans are almost certainly the greatest species to ever walk the planet Earth—and the universe as far as we can tell—but at the end of the day Mother Nature still completely dictates our lives. To destroy her is to destroy ourselves.

In 1964, Ronald Reagan (at the time governor of California) said something during a speech that, at the time, was about the Vietnam War, but can now apply (perhaps even more accurately now than back then) to the current climate situation:

> "We are at war with the most dangerous enemy that has ever faced mankind in his long climb from the swamp to the stars, and it has been said if we lose that war, and in doing so lose this way of freedom of ours, history will record with the greatest astonishment that those who had the most to lose did the least to prevent its happening."

Notes

A Note on Notes: Many of the following sources are subject to change over time. Therefore, when accessed, information may have changed from what is reported in the text of this book. Some, but not all, of the sources provide historical data that can be viewed which should be discoverable with the information provided in these notes.

Introduction

1   Cohen, Jennie. "Human Ancestors Tamed Fire Earlier Than Thought." *History.com*, A&E Television Networks, 22 Aug. 2018, www.history.com/news/human-ancestors-tamed-fire-earlier-than-thought.

2   Arrhenius, Svante. "On the Influence of Carbonic Acid in the Air upon the Temperature of the Ground." *Philosophical Magazine and Journal of Science*, vol. 41, no. 251, ser. 5, Apr. 1896, pp. 237–276. *5*. (Page 268)

3   Sabas, Matthew. "History of Solar Power." *IER*, 18 Feb. 2016, www.instituteforenergyresearch.org/renewable/solar/history-of-solar-power/.

4   As is discussed in later sections of this book, the IER is not always the most reputable source. Having received significant funding from fossil fuel interests, it frequently engages in questionable reporting. In saying that, this article has all of its sources cited and provides a good overview of solar powers history in the US.

5   "United States Surpasses 2 Million Solar Installations." *Solar Energy Industries Association (SEIA)*, 9 May 2019, www.seia.org/news/united-states-surpasses-2-million-solar-installations.

6   IPCC, 2014: Climate Change 2014: Synthesis Report. Contribution of Working Groups I, II and III to the Fifth Assessment Report of the Intergovernmental Panel on Climate Change [Core Writing Team, R.K. Pachauri and L.A. Meyer (eds.)]. IPCC, Geneva, Switzerland

Chapter 2: When the Sun Shines

1   Creighton, Jolene. "The Kardashev Scale - Type I, II, III, IV & V Civilization." *Futurism*, 19 July 2014, futurism.com/the-kardashev-scale-type-i-ii-iii-iv-v-civilization.

2   Tsao, Jeff, et al. "The Sun Delivers More Energy to Earth in an Hour than We Use in a Year from Fossil, Nuclear and All Renewable Sources Combined." *Sandia National Laboratories*, U.S. Department of Energy, 20 Apr. 2006, www.sandia.gov/~jytsao/Solar%20FAQs.pdf.

3   This is worth putting in context. If the sun delivers all of our energy in 2 hours, that means if we covered a proportionate amount of land area to the amount of time the sun power the earth, we could power the earth. Two hours is about 0.02% of the year. 0.02% of the Earth's land surface area is about 150 million km2, meaning a solar world would require around 3.4 million km2, an area almost equal to 35% of the US surface area. This is assuming the power is always consistent. It would actually need to be double this, to account for day/night cycles. This is obviously also ignoring massive land requirement implications such as infrastructure, transmission distance, panel spacing, and the land requirements from mining, distribution, workers offices etc.

4   "U.S. Energy Information Administration - EIA - Independent Statistics and Analysis." *U.S. Energy Facts Explained - Consumption and Production - U.S. Energy Information Administration (EIA)*, 28 Aug. 2019, www.eia.gov/energyexplained/us-energy-facts/.

5   And, in-fact, you can't line every rooftop with solar panels. Many houses' roofs are simply not built to be able to hold the additional weight and need reinforcement or completely redone to fit this accommodation.

6   "Cities Are at the Frontline of the Energy Transition." *IEA*, 7 Sept. 2016, www.iea.org/news/cities-are-at-the-frontline-of-the-energy-transition.

7   Penin, Ivan. "California Invested Heavily in Solar Power. Now There's so Much That Other States Are Sometimes Paid to Take It." *Los Angeles Times*, 22 June 2017, www.latimes.com/projects/la-fi-electricity-solar/.

8   "California Imports the Most Electricity from Other States; Pennsylvania Exports the Most." *U.S. Energy Information Administration (EIA)*, 4 Apr. 2019, www.eia.gov/todayinenergy/detail.php?id=38912.

9   "State Electricity Profiles (Data for 2019)." *Energy Information Administration*, 31 Dec. 2019, www.eia.gov/electricity/state/.

10   Ibid

11   "Electric Power Monthly." *U.S. Energy Information Administration (EIA)*, www.eia.gov/electricity/monthly/epm_table_grapher.php?t=epmt_5_6_a.

12   Ibid

13   "Net Public Electricity Generation in Germany in 2019." *Electricity Generation | Energy Charts*, Fraunhofer ISE, 12 Jan. 2020, 20:59, www.energy-charts.de/energy_pie.htm?year=2019.

14   "Germany." *Environmental Progress*, 3 Sept. 2019, environmentalprogress.org/germany/.

15   Shellenberger, Michael, director. *Why Renewables Can't Save the Planet. YouTube*, TEDx Talks, 4 Jan. 2019,www.youtube.com/watch?v=N-yALPEpV4w&list=PLI0ToICTHWJGvDTEs8rjNstI2m_p7IXxl&index=17&t=900s.

16   Desai, Jemin, and Mark Nelson. "Are We Headed for a Solar Waste Crisis?" *Environmental Progress*, 21 June 2017, environmentalprogress.org/big-news/2017/6/21/are-we-headed-for-a-solar-waste-crisis.

17   "End-of-Life Management Solar Photovoltaic Panels." *International Renewable Energy Agency (IRENA)*, June 2016, www.irena.org/publications/2016/Jun/End-of-life-management-Solar-Photovoltaic-Panels. ISBN : 978-92-95111-99-8

18   "Carbon Nanotubes Might Be the Secret Boost Solar Energy Has Been Looking For." *Seeker*, Youtube, 16 Sept. 2019. www.youtube.com/watch?v=EwiDGxkD9_c&list=PLI0ToICTHWJGvDTEs8rjNstI2m_p7IXxl&index=3&t=24s

19   IEA (2020), Solar PV, IEA, Paris https://www.iea.org/reports/solar-pv (Pulled from graph "Net solar PV capacity additions, 2017-2019")

Chapter 3: When the Wind Blows

1   "Denmark - Key Energy Statistics 2018." *IEA*, United Nations, 2020, www.iea.org/countries/denmark.

2   IEA (2020), Renewables 2020, IEA, Paris https://www.iea.org/reports/renewables-2020 (Pulled from graphs "Onshore wind net capacity additions by country or region, 2015-2022" and "Offshore wind net capacity additions by country or region, 2016-2022")

3   Ibid (Pulled from graph "China quarterly wind deployment, 2018-2020", comparison made to graph "Onshore wind net capacity additions by country or region, 2015-2022")

4   Ibid (Pulled from graph "US wind project development status, 2017-2022")

5   Ibid (Pulled from graphs "Onshore wind net capacity additions by country or region, 2015-2022" and "Offshore wind net capacity additions by country or region, 2016-2022")

6   "Electricity Generation, Capacity, and Sales in the United States." *U.S. Energy Information Administration (EIA)*, 19 Mar. 2020, www.eia.gov/energyexplained/electricity/electricity-in-the-us-generation-capacity-and-sales.php#:~:text=At%20the%20end%20of%202019,%20the%20United%20States,kW%20of%20small-scale%20solar%20photovoltaic%20electricity%20generating%20capacity.

7   Feldman, Sarah. "Wind Turbines Are Not Killing Fields for Birds." *Statista*, 3 Sept. 2019, www.statista.com/chart/15195/wind-turbines-are-not-killing-fields-for-birds/.

8   "Protect Eagles from Wind Turbine Fatalities." *American Eagle Foundation*, www.eagles.org/take-action/wind-turbine-fatalities/.

9   "Wind Power Getting Headwind in Germany." *Deutsche Welle (DW) Documentary*, Youtube, 11 Aug. 2020.

10   Berwyn, Bob. "How Do Offshore Wind Farms Affect Ocean Ecosystems?: DW: 22.11.2017." *DW.COM*, 22 Nov. 2017, www.dw.com/en/how-do-offshore-wind-farms-affect-ocean-ecosystems/a-40969339.

11   "1 Kilowatt-Hour." *BlueSkyModel*, blueskymodel.org/kilowatt-hour.

12   "Environmental Impacts of Wind Power." *Union of Concerned Scientists*, 5 Mar. 2013, www.ucsusa.org/resources/environmental-impacts-wind-power

13   Nugent, Daniel, and Benjamin K. Sovacool. "Assessing the Lifecycle Greenhouse Gas Emissions from Solar PV and Wind Energy: A Critical Meta-Survey." *Energy Policy Vol. 65 Pages 229-244*, Elsevier, Feb. 2014, www.sciencedirect.com/science/article/pii/S0301421513010719.

14   "1 Kilowatt-Hour." *BlueSkyModel*, blueskymodel.org/kilowatt-hour.

15   "Products & Applications of Sulphur Hexafluoride (SF6)." *Fluorocarbons*, EFCTC, www.fluorocarbons.org/products-applications-sulphur-hexafluoride-sf6/#.XioVhGhKiUk.

16   "Sulphur Hexafluoride." *Pollutant Fact Sheet*, Scottish Environment Protection Agency (SEPA), apps.sepa.org.uk/spripa/Pages/SubstanceInformation.aspx?pid=10.

17   McGrath, Matt. "Climate Change: Electrical Industry's 'Dirty Secret' Boosts Warming." *BBC News*, BBC, 13 Sept. 2019, www.bbc.com/news/science-environment-49567197?ref= WindEurope.

18   "Wind Energy and SF6 in Perspective." *WindEurope*, 25 Sept. 2019, windeurope.org/newsroom/news/wind-energy-and-sf6-in-perspective/.

19   IEA (2020), Offshore Wind, IEA, Paris https://www.iea.org/reports/offshore-wind (Pulled from graph "Offshore wind power generation in the Sustainable Development Scenario")

20   IEA (2019), Offshore Wind Outlook 2019, IEA, Paris https://www.iea.org/reports/offshore-wind-outlook-2019

21   Ibid (Pulled from graph "Installed offshore wind capacity, 2018 and 2040, Stated Policies Scenario")

22   Ibid

23   Roberto Lacal-Arántegui, José M. Yusta, José Antonio Domínguez-Navarro, Offshore wind installation: Analysing the evidence behind improvements in installation time, Renewable and Sustainable Energy Reviews, Volume 92, 2018, Pages 133-145, ISSN 1364-0321, https://doi.org/10.1016/j.rser.2018.04.044.

24   "U.S. Wind Power Achieves Landmark Installed Capacity of 82,000 Megawatts." *U.S. Department of Energy (DOE)*, 8 Feb. 2017, www.energy.gov/eere/wind/articles/us-wind-power-achieves-landmark-installed-capacity-82000-megawatts.

25   "How Do Wind Turbines Survive Severe Storms?" *U.S. Department of Energy (DOE)*, www.energy.gov/eere/articles/how-do-wind-turbines-survive-severe-storms.

Chapter 4: Where the Water Runs

1   "Hydropower Explained." *EIA*, U.S. Energy Information Administration , 12 Apr. 2019, www.eia.gov/energyexplained/hydropower/.

2   "Hydroelectric Power Water Use." *Usgs*, United States Geological Survey, www.usgs.gov/special-topic/water-science-school/science/hydroelectric-power-water-use?qt-science_center_objects=0#qt-science_center_objects.

3   "Brazil - Key Energy Statistics 2018; Energy Topic- Energy Supply, Indicator- Electricity Generation by Source" *IEA*, 11 Dec. 2019, www.iea.org/countries/Brazil.

4   *Hydroelectric Power*. Bureau of Reclamation, 2005, *Hydroelectric Power*, www.usbr.gov/power/edu/pamphlet.pdf Page 2.

5   "Benefits of Dams." *FEMA*, Department of Homeland Security, 22 Oct. 2019, 09:55, www.fema.gov/benefits-dams.

6   "Safety of Nuclear Power Reactors." *World Nuclear Association*, June 2019, www.world-nuclear.org/information-library/safety-and-security/safety-of-plants/safety-of-nuclear-power-reactors.aspx.

7   (Also see) Hirschberg, S., Speikerman, G., Dones, R. *"Severe Accidents in the Energy Sector."* Swiss Federal Office of Energy, 1998, *Severe Accidents in the Energy Sector*, inis.iaea.org/collection/NCLCollectionStore/_Public/30/045/30045581.pdf.

8   Usigbe, Leon. "Drying Lake Chad Basin Gives Rise to Crisis." *Africa Renewal*, United Nations, 24 Dec. 2019, www.un.org/africarenewal/magazine/december-2019-march-2020/drying-lake-chad-basin-gives-rise-crisis.

9   Scott, Rebecca. "Why Is Lake Chad Drying up?" *The Glasgow Guardian*, 1 Mar. 2020, glasgowguardian.co.uk/2020/03/01/why-is-lake-chad-drying-out/.

10   "Hydropower and the Environment." *EIA-Independent Statistics and Analysis*, U.S. Energy Information Administration, 13 Nov. 2019, www.eia.gov/energyexplained/hydropower/hydropower-and-the-environment.php.

11   "Understanding Global Warming Potentials." *EPA*, Environmental Protection Agency, 14 Feb. 2017, www.epa.gov/ghgemissions/understanding-global-warming-potentials.

12   Ivan B.T. Lima, Fernando M. Ramos, Luis A. W. Bambace, Reinaldo R. Rosa. "Methane Emissions from Large Dams as Renewable Energy Resources: A Developing Nation Perspective," Mitigation and Adaptation Strategies for Global Change, March 2007. https://link.springer.com/article/10.1007/s11027-007-9086-5

13   "4% Of Global Warming Due to Dams, Says New Research." *International Rivers*, 9 May 2007, www.internationalrivers.org/resources/4-of-global-warming-due-to-dams-says-new-research-3868.

14 Crettenand, Nicholas; Finger, Matthias; Puttgen, Hans Björn (2012). "The facilitation of mini and small hydropower in Switzerland: shaping the institutional framework. With a particular focus on storage and pumped-storage schemes". *Ecole Polytechnique Fédérale de Lausanne (EPFL). PhD Thesis N° 5356.* https://infoscience.epfl.ch/record/176337?ln=en

15 "Salinity Gradient." *Ocean Energy Europe,* www.oceanenergy-europe.eu/ocean-energy/salinity-gradient/.

16 Auffret, Jerome-Cecil. *Dynamic Salt. Curiosity Stream,* 2015, curiositystream.com/video/1914/dynamic-salt. 18:30

17 "Tidal Power." *EIA-Independent Statistics and Analysis,* U.S. Energy Information Administration, 23 Sept. 2019, www.eia.gov/energyexplained/hydropower/tidal-power.php.

18 "Three Gorges Dam: The World's Largest Hydroelectric Plant." *U.S. Geological Survey (USGS),* www.usgs.gov/special-topic/water-science-school/science/three-gorges-dam-worlds-largest-hydroelectric-plant?qt-science_center_objects=0#qt-science_center_objects.

Chapter 5: Where the Earth Is Hot

1   IEA (2020), Geothermal, IEA, Paris https://www.iea.org/reports/geothermal (Pulled from graph "Geothermal power generation in the Sustainable Development Scenario, 2000-2030")

2   Unwin, Jack. "Larderello - the Oldest Geothermal Power Plant in the World." *Power Technology | Energy News and Market Analysis*, 8 Oct. 2019, www.power-technology.com/features/oldest-geothermal-plant-larderello/.

3   "Geothermal FAQs." *Energy.gov*, US Department of Energy, www.energy.gov/eere/geothermal/geothermal-faqs#cost_to_develop_geothermal_power_plant.

4   *Levelized Cost and Levelized Avoided Cost of New Generation Resources in the Annual Energy Outlook 2021*. US Energy Information Agency. February, 2021, www.eia.gov/outlooks/aeo/pdf/electricity_generation.pdf (Page 9)

5   Harmon, Katherine. "How Does Geothermal Drilling Trigger Earthquakes?" *Scientific American*, Scientific American, 29 June 2009, www.scientificamerican.com/article/geothermal-drilling-earthquakes/.

6   Gabbatt, Adam. "Swiss Geothermal Power Plan Abandoned after Quakes Hit Basel." *The Guardian*, Guardian News and Media, 15 Dec. 2009, www.theguardian.com/world/2009/dec/15/swiss-geothermal-power-earthquakes-basel.

7   "Geothermal Explained-Where Geothermal Energy Is Found." *EIA- Independent Statistics and Analysis*, U.S. Energy Information Administration, 5 Dec. 2019, www.eia.gov/energyexplained/geothermal/where-geothermal-energy-is-found.php.

8   "Geothermal Explained- Use of geothermal energy" *U.S. Energy Information Administration - EIA*, 5 Dec. 2019, www.eia.gov/energyexplained/geothermal/where-geothermal-energy-is-found.php.

9   National Geographic Society. "Ring of Fire." *National Geographic Society*, 4 Apr. 2019, www.nationalgeographic.org/encyclopedia/ring-fire/.

10   IEA (2020), Kenya, IEA, Paris Kenya - Countries & Regions - IEA (Pulled from graph "Electricity generation by source, Kenya 1990-2018" using filters Electricity and heat—under energy topic—and Electricity generation by source—under indicator.)

11   IEA (2020), Iceland, IEA, Paris Kenya - Countries & Regions - IEA (Pulled from graph "Electricity generation by source, Iceland 1990-2018" using filters Electricity and heat—under energy topic—and Electricity generation by source—under indicator.)

12   "Kola Superdeep Borehole (KSDB) - IGCP 408: Rocks and Minerals at Great Depths and on the Surface." *Internet Archives*, ICDP:Kola, 7 May 2014, web.archive.org/web/20140507113717/www-icdp.icdp-online.org/front_content.php?idcat=695.

13   Ibid

14   "Geothermal Explained: Use of geothermal energy" *U.S. Energy Information Administration (EIA)*, 22 March 2021, www.eia.gov/energyexplained/geothermal/where-geothermal-energy-is-found.php.

Chapter 6: The Solution We Hate

1   Reinhart, RJ. "40 Years After Three Mile Island, Americans Split on Nuclear Power." *Gallup.com*, Gallup, 27 Mar. 2019, news.gallup.com/poll/248048/years-three-mile-island-americans-split-nuclear-power.aspx.

2   "Americans' Opinion on Renewables and Other Energy Sources." *Pew Research Center Science & Society*, Pew Research Center, 4 Oct. 2016, www.pewresearch.org/science/2016/10/04/public-opinion-on-renewables-and-other-energy-sources/.

3   "Generation 3 Nuclear Reactors." *IAEA*, CEA, 2010, inis.iaea.org/collection/NCLCollection Store/_Public/44/078/44078364.pdf. (Page 3)

4   "Nuclear Energy Factsheet." *Center for Sustainable Systems*, University of Michigan, css.umich.edu/factsheets/nuclear-energy-factsheet.

5   "Nuclear Costs in Context." *Nuclear Energy Institute (NEI)*, Oct. 2020, www.nei.org/resources/reports-briefs/nuclear-costs-in-context. (Page 9)

6   "Jobs." *Nuclear Energy Institute (NEI)*, www.nei.org/advantages/jobs.

7   "Fossil Fuel Industry Climate Science Deception." *Union of Concerned Scientists*, 29 Mar. 2016, www.ucsusa.org/about/news/fossil-fuel-industry-climate-science-deception.

8   "Tweet the Story of the Fossil Fuel Industry's Climate Deception." *Union of Concerned Scientists*, 11 Mar. 2017, www.ucsusa.org/resources/tweet-story-fossil-fuel-industrys-climate-deception.

9   "Holding Major Fossil Fuel Companies Accountable for Nearly 40 Years of Climate Deception and Harm." *Union of Concerned Scientists*, 22 Mar. 2016, www.ucsusa.org/resources/holding-major-fossil-fuel-companies-accountable.

10   "The War on Nuclear." *Environmental Progress*, environmentalprogress.org/the-war-on-nuclear.

11   Conca, James. "Natural Gas Industry Blasts Nuclear Power With Fake News." *Forbes*, 15 June 2017, www.forbes.com/sites/jamesconca/2017/06/15/natural-gas-industry-blasts-nuclear-power-with-fake-news/?sh=536a0a86133b.

12   Silverstein, Ken. "Are Fossil Fuel Interests Bankrolling The Anti-Nuclear Energy Movement?" *Forbes*, 13 July 2016, www.forbes.com/sites/kensilverstein/2016/07/13/are-fossil-fuel-interests-bankrolling-the-anti-nuclear-energy-movement/?sh=57ea2e307453.

13   Adams, Rod. "Smoking Gun Archives." *Atomic Insights*, 21 May 2020, atomicinsights.com/smoking-gun/.

14   Kuzoian, Alex. "Bananas Give You More Radiation Exposure than Living next to a Nuclear Power Plant." *Business Insider*, 9 July 2017, www.businessinsider.com.au/how-much-radiation-youre-exposed-to-in-everyday-life-2017-7.

Chapter 7: The Great Misconception

1   Afework, Bethel; Hanania, Jordan; Stenhouse, Kailyn and Jason Donev. "Control Rod." *Energy Education*, University of Calgary, 18 May 2018, energyeducation.ca/encyclopedia/Control_rod.

2   "Water Cooled Reactors." *International Atomic Energy Agency (IAEA)*, www.iaea.org/topics/water-cooled-reactors.

3   Matson, John. "What Happens During a Nuclear Meltdown?" *Scientific American*, Scientific American, 15 Mar. 2011, www.scientificamerican.com/article/nuclear-energy-primer/.

4   "Safety of Nuclear Power Reactors." *World Nuclear Association*, June 2019, www.world-nuclear.org/information-library/safety-and-security/safety-of-plants/safety-of-nuclear-power-reactors.aspx.

5   This figure is not technically correct, but it is close enough. As is discussed later in this book, one person died from radiation during the Fukushima cleanup and another incident in the US resulted in the death of a couple radiation workers. For all extents and purposes, the previous quote is accurate.

6   "Heart Disease Facts." *Centers for Disease Control and Prevention*, 2 Dec. 2019, www.cdc.gov/heartdisease/facts.htm.

7   "Air Pollution and Cardiovascular Disease." *American Heart Association, American Stroke Association*, 2014, www.heart.org/idc/groups/heart-public/@wcm/@adv/documents/downloadable/ucm_463344.pdf.

8   Pushker A. Kharecha and James E. Hansen. *Environmental Science & Technology* 2013 47 (9), 4889-4895 DOI: 10.1021/es3051197

9   (Also see) Kharecha, Pushker, and James Hansen. "Coal and Gas Are Far More Harmful than Nuclear Power." *NASA-Global Climate Change*, NASA, 22 Apr. 2013, climate.nasa.gov/news/903/coal-and-gas-are-far-more-harmful-than-nuclear-power/.

10   "Radioactive Waste Management." *World Nuclear Association*, Feb. 2020, world-nuclear.org/information-library/nuclear-fuel-cycle/nuclear-wastes/radioactive-waste-management.aspx.

11   Ibid

12   Ibid

13   Ibid

14   *Status and Trends in Spent Fuel and Radioactive Waste Management.* 2018, *International Atomic Energy Agency*, United Nations, www-pub.iaea.org/MTCD/Publications/PDF/P1799_web.pdf. FIG. 22 (Pages 38-39)

15   "Radioactive Waste Management." *World Nuclear Association*, Apr. 2018, www.world-nuclear.org/information-library/nuclear-fuel-cycle/nuclear-wastes/radioactive-waste-management.aspx

16   "Nuclear Waste." *Nuclear Energy Institute (NEI)*, www.nei.org/fundamentals/nuclear-waste.

17   "Electricity Information 2019 – Analysis." *IEA*, Sept. 2019, www.iea.org/reports/electricity-information-2019. (Pulled from graph "World gross electricity production, by source, 2018")

18   McNeely, E., Mordukhovich, I., Staffa, S. *et al.* Cancer prevalence among flight attendants compared to the general population. *Environ Health* 17, 49 (2018). https://doi.org/10.1186/s12940-018-0396-8

19   Lynch, Patrick. "Modeling Radiation Exposure for Pilots, Crew and Passengers on Commercial Flights." *NASA*, 19 Dec. 2008, www.nasa.gov/topics/earth/features/AGU-NAI-RAS.html#:~:text=While%20it%20may%20not%20be%20commonly%20known%2C%20airline,on%20an%20annual%20basis%20than%20nuclear%20plant%20workers.

20   Hvistendahl, Mara. "Coal Ash Is More Radioactive Than Nuclear Waste." *Scientific American*, 13 Dec. 2007, www.scientificamerican.com/article/coal-ash-is-more-radioactive-than-nuclear-waste/.

21   Mueller, Mike. "7 Things The Simpsons Got Wrong About Nuclear." *U.S. Department of Energy (DOE)*, 4 Apr. 2014, www.energy.gov/ne/articles/7-things-simpsons-got-wrong-about-nuclear.

22   "Radioactive Waste Management." *World Nuclear Association*, Apr. 2018, www.world-nuclear.org/information-library/nuclear-fuel-cycle/nuclear-wastes/radioactive-waste-management.aspx.

23   "The Legacy of U.S. Nuclear Testing and Radiation Exposure in the Marshall Islands." *U.S. Embassy in the Republic of the Marshall Islands*, 15 Sept. 2012, mh.usembassy.gov/the-legacy-of-u-s-nuclear-testing-and-radiation-exposure-in-the-marshall-islands/#:~:text=The%20United%20States%20conducted%2067%20nuclear%20explosive%20tests,and%20ended%20all%20nuclear%20explosive%20testing%20in%201992.

24   Rust, Susanne. "How the U.S. Betrayed the Marshall Islands, Kindling the next Nuclear Disaster." *Los Angeles Times*, 10 Nov. 2019, www.latimes.com/projects/marshall-islands-nuclear-testing-sea-level-rise/.

25   Ibid

26   Brugge, Doug, et al. "The American Journal of Public Health (AJPH) from the American Public Health Association (APHA) Publications." *American Public Health Association (APHA)*, 10 Oct. 2011, ajph.aphapublications.org/doi/10.2105/AJPH.2006.103044#_i2.

27   Ibid

28   Gilliland, Frank D. MD; Hunt, William C. MS; Pardilla, Marla MSW, MPH; Key, Charles R. MD, PhD Uranium Mining and Lung Cancer Among Navajo Men in New Mexico and Arizona, 1969 to 1993, Journal of Occupational and Environmental Medicine: March 2000 - Volume 42 - Issue 3 - p 278-283 (https://journals.lww.com/joem/pages/articleviewer.aspx?year=2000&issue=03000&article=00008&type=abstract#:~:text=Uranium%20Mining%20and%20Lung%20Cancer%20Among%20Navajo%20Men,risk%20of%20lung%20cancer%20for%20Navajo%20uranium%20miners.)

29   Ibid

30   "World Uranium Mining Production." *World Nuclear Association*, Dec. 2020, www.world-nuclear.org/information-library/nuclear-fuel-cycle/mining-of-uranium/world-uranium-mining-production.aspx.

31   "Uranium Mining Overview." *World Nuclear Association*, Dec. 2020, www.world-nuclear.org/information-library/nuclear-fuel-cycle/mining-of-uranium/uranium-mining-overview.aspx.

32   "Country Nuclear Power Profiles-France." *IAEA*, United Nations, 2019, cnpp.iaea.org/countryprofiles/France/France.htm.

33   "How Much Does Electricity Cost in France?" *Selectra*, en.selectra.info/energy-france/guides/electricity/cost.

34   Nelson, Mark, and Madison Czerwinski. "With Nuclear Instead of Renewables, California & Germany Would Already Have 100% Clean Electricity." *Environmental Progress*, 11 Sept. 2018, environmentalprogress.org/big-news/2018/9/11/california-and-germany-decarbonization-with-alternative-energy-investments.

35   Staudenmaier, Rebecca. "Germany's Nuclear Phaseout Explained." *Deutsche Welle (DW)*, 15 June 2017, www.dw.com/en/germanys-nuclear-phaseout-explained/a-39171204.

36   IEA (2020), Germany 2020, IEA, Paris https://www.iea.org/reports/germany-2020

37   Watts, Nick, et al. "The 2020 Report of The Lancet Countdown on Health and Climate Change: Responding to Converging Crises." *The Lancet*, vol. 397, no. 10269, 2021, pp. 129–170., doi:10.1016/s0140-6736(20)32290-x. https://www.thelancet.com/infographics/countdown2020 (Pulled from map "Premature Deaths from Ambient Air Pollution". Using data found online, Germanys 2018 population was assumed to be 83 million and Frances 2018 population was assumed to be 65 million.)

38   "Should We Go Nuclear?" *How to Save a Planet*, Gimlet Media, Spotify, Dec. 2020. This podcast episode does a good job of explaining part of the reasoning behind this and provides several good examples of major cost overruns. For other information, just searching nuclear power cost overruns on the internet will give you a plethora of examples.

39   "Capital Cost Estimates for Utility Scale Electricity Generating Plants." *U.S. Energy Information Agency (EIA)*, Nov. 2016, www.eia.gov/analysis/studies/powerplants/capitalcost/pdf/capcost_assumption.pdf. (Page 7)

40   J.M.K.C. Donev et al. (2018). Energy Education - Energy density [Online]. Available: https://energyeducation.ca/encyclopedia/Energy_density. [Accessed: March 23, 2021].

41   Nalbandian-Sugden, Herminé. "Operating Ratio and Cost of Coal Power Generation." *IEA Clean Coal Centre*, International Energy Agency (IEA), Dec. 2016, usea.org/sites/default/files/Operating%20ratio%20and%20cost%20of%20coal%20power%20generation%20-%20ccc272-1.pdf. IEACCC Ref: CCC/272 ISBN: 978–92–9029–595-2 (Page 33)

42   "Nuclear Costs in Context." *Nuclear Energy Institute (NEI)*, Oct. 2020, www.nei.org/resources/reports-briefs/nuclear-costs-in-context. (Page 6)

43   "Nuclear Power in China." *World Nuclear Association*, Jan. 2021, www.world-nuclear.org/information-library/country-profiles/countries-a-f/china-nuclear-power.aspx. (Pulled from graph "Operable nuclear power capacity")

44   "2035 Report." *Goldman School of Public Policy*, University of California Berkley, June 2020. (Page 20)

45   IEA (2020), Nuclear Power, IEA, Paris https://www.iea.org/reports/nuclear-power

46   Ibid (Pulled from graph "Global nuclear capacity by scenario, 2000-2040")

47   Ibid

Chapter 8: The Legacy of Incompetence

1   Semenov, B. A. "Nuclear Power in the Soviet Union." *International Atomic Energy Agency (IAEA)*, United Nations, www.iaea.org/sites/default/files/25204744759.pdf.

2   "Backgrounder on the Three Mile Island Accident." *United States Nuclear Regulatory Commission*, 21 June 2018, www.nrc.gov/reading-rm/doc-collections/fact-sheets/3mile-isle.html.

3   "Lessons From the 1979 Accident at Three Mile Island." *Nuclear Energy Institute (NEI)*, Oct. 2019, nei.org/resources/fact-sheets/lessons-from-1979-accident-at-three-mile-island.

4   "Three Mile Island Accident." *World Nuclear Association*, Mar. 2020, world-nuclear.org/information-library/safety-and-security/safety-of-plants/three-mile-island-accident.aspx.

5   Ibid

6   "Radiation Dose Calculator." *American Nuclear Society (ANS)*, ans.org/pi/resources/dosechart/msv.php.

7   "The Effects of the Three Mile Island Accident Meltdown After 40 Years." *UC Press Blog*, University of California, 28 Mar. 2019, www.ucpress.edu/blog/42233/the-effects-of-the-three-mile-island-accident-meltdown-after-40-years/

8   Study Citation: Talbott, Evelyn O, et al. "Long-Term Follow-up of the Residents of the Three Mile Island Accident Area: 1979-1998." *Environmental Health Perspectives*, vol. 111, no. 3, 2003, pp. 341–348., doi:10.1289/ehp.5662.

9   "Chernobyl Accident 1986." *World Nuclear Association*, June 2019, www.world-nuclear.org/information-library/safety-and-security/safety-of-plants/chernobyl-accident.aspx.

10   International Atomic Energy Agency, The International Chernobyl Project, IAEA, Vienna (1991)

11   "Chernobyl's Legacy: Health, Environmental and Socio-Economic Impacts and Recommendations to the Governments of Belarus, the Russian Federation and Ukraine." *The Chernobyl Forum*, International Atomic Energy Agency (IAEA), www.iaea.org/sites/default/files/chernobyl.pdf. (Page 22)

12   "Chernobyl: The True Scale of the Accident." *IAEA*, United Nations, 5 Sept. 2005, www.iaea.org/newscenter/pressreleases/chernobyl-true-scale-accident.

13   Lochbaum, Dave. "Nuclear Plant Containment Failure: Robusted." *All Things Nuclear*, Union of Concerned Scientists, 30 May 2016, allthingsnuclear.org/dlochbaum/nuclear-plant-containment-failure-robusted.

14   "Frequently Asked Chernobyl Questions." *IAEA*, United Nations, 7 Nov. 2016, www.iaea.org/newscenter/focus/chernobyl/faqs.

15   "The Great Wave of Reform." *NASA*, 8 Jan. 2020, nsc.nasa.gov/resources/case-studies/detail/the-great-wave-of-reform.

16   "FAQs: Health Consequences of Fukushima Daiichi Nuclear Power Plant Accident in 2011." *World Health Organization*, United Nations, Aug. 2015, www.who.int/ionizing_radiation/a_e/fukushima/faqs-fukushima/en/.

17   "Quake Moved Japan by 8 Feet: USGS." *Phys.org*, 13 Mar. 2011, phys.org/news/2011-03-quake-japan-feet-usgs.html.

18   Dunbar, Brian. "Japan Quake May Have Shortened Earth Days, Moved Axis." *NASA*, 14 Mar. 2011, www.nasa.gov/topics/earth/features/japanquake/earth20110314.html#:~:text=The%20March%202011,%202011,%20great%20earthquake%20in%20Japan,axis.%20But%20don't%20worry%E2%80%94you%20won't%20notice%20the%20difference.

19   "Fukushima Daiichi Accident." *World Nuclear Association*, Oct. 2018, www.world-nuclear.org/information-library/safety-and-security/safety-of-plants/fukushima-accident.aspx.

20   Biello, David. "Partial Meltdowns Led to Hydrogen Explosions at Fukushima Nuclear Power Plant." *Scientific American*, 14 Mar. 2011, www.scientificamerican.com/article/partial-meltdowns-hydrogen-explosions-at-fukushima-nuclear-power-plant/.

21   "Fukushima Daiichi Accident." *World Nuclear Association*, Oct. 2018, www.world-nuclear.org/information-library/safety-and-security/safety-of-plants/fukushima-accident.aspx.

22   Kumerasan J. "Preparedness and resilience: the hallmarks of response and recovery." *Western Pacific Surveillance and Response Journal*, 2011, 2(3):1-2. doi:10.5365/wpsar.year.2011.2.4.013

23   "FAQs: Health Consequences of Fukushima Daiichi Nuclear Power Plant Accident in 2011." *World Health Organization*, United Nations, Aug. 2015, www.who.int/ionizing_radiation/a_e/fukushima/faqs-fukushima/en/.

24   "Japan Confirms First Fukushima Worker Death from Radiation." *BBC News*, 5 Sept. 2018, www.bbc.com/news/world-asia-45423575.

25   "Fukushima: Radiation Exposure." *World Nuclear Association*, Feb. 2016, www.world-nuclear.org/information-library/safety-and-security/safety-of-plants/appendices/fukushima-radiation-exposure.aspx.

26   "Chernobyl's Legacy: Health, Environmental and Socio-Economic Impacts and Recommendations to the Governments of Belarus, the Russian Federation and Ukraine." *The Chernobyl Forum*, International Atomic Energy Agency (IAEA), www.iaea.org/sites/default/files/chernobyl.pdf.

27   Steinhauser, Georg, et al. "Comparison of the Chernobyl and Fukushima Nuclear Accidents: A Review of the Environmental Impacts." *Science of The Total Environment*, vol. 470-471, 2014, pp. 800–817., doi: 10.1016/j.scitotenv.2013.10.029.

28   "The Fukushima Daiichi Accident: Technical Volume 1, Description and Context of the Accident." *International Atomic Energy Agency (IAEA)*, 2015, www-pub.iaea.org/MTCD/Publications/PDF/AdditionalVolumes/P1710/Pub1710-TV1-Web.pdf. (Page 150, figures 1.4-4 & 1.4-5.)

29   "Radioisotope Brief: Cesium-137 (Cs-137)." *U.S. Centers for Disease Control and Prevention (CDC)*, 4 Apr. 2018, www.cdc.gov/nceh/radiation/emergencies/isotopes/cesium.htm.

30   Little, Jane Braxton. "Fukushima Residents Return Despite Radiation." *Scientific American*, 16 Jan. 2019, www.scientificamerican.com/article/fukushima-residents-return-despite-radiation/.

31   Woodyatt, Amy. "Wildlife Flourishing in Uninhabited Areas around Fukushima." *CNN*, 8 Jan. 2020, www.cnn.com/2020/01/07/asia/fukushima-wildlife-intl-scli-scn/index.html.

32   "Motor Vehicle Crash Deaths." *Centers for Disease Control and Prevention (CDC)*, 6 July 2016, www.cdc.gov/vitalsigns/motor-vehicle-safety/index.html.

33   "Current Cigarette Smoking Among Adults in the United States." *Centers for Disease Control and Prevention (CDC)*, 10 Dec. 2020, www.cdc.gov/tobacco/data_statistics/fact_sheets/adult_data/cig_smoking/index.htm.

34   "Why Does Air Matter?" *United Nations Environment Programme (UNEP)*, www.unep.org/explore-topics/air/why-does-air-matter.

Chapter 9: Thirty More Years...

1   "Space Technology 5." *Jet Propulsion Lab (JPL)*, NASA, Jan. 2004, www.jpl.nasa.gov/nmp/st5/SCIENCE/sun.html.

2   "Ivy Mike, 1 November 1952 - First Full-Scale Thermonuclear Test." *CTBTO Preparatory Commission*, www.ctbto.org/specials/testing-times/1-november-1952-ivy-mike/.

3   "Big Ivan, The Tsar Bomba ('King of Bombs')." *Nuclear Weapon Archive*, 3 Sept. 2007, nuclearweaponarchive.org/Russia/TsarBomba.html.

4   Ibid

5   Ibid

6   Ibid

7   "60 Years of Progress- Milestones around the World." *ITER*, www.iter.org/sci/BeyondITER.

8   "ITER Members." *ITER*, www.iter.org/proj/Countries.

9   "What Is ITER?" *ITER*, www.iter.org/proj/inafewlines.

10   "After ITER." *ITER*, www.iter.org/sci/iterandbeyond.

11   "Facts & Figures." *ITER*, www.iter.org/FactsFigures.

12   "Plasma Confinement." *ITER*, www.iter.org/sci/PlasmaConfinement.

13   "Advantages of Fusion." *ITER*, www.iter.org/sci/Fusion.

14   Craig Freudenrich, Ph.D., and John Fuller. "How Nuclear Bombs Work." *HowStuffWorks Science*, 27 Jan. 2020, science.howstuffworks.com/nuclear-bomb6.htm.

15   "Fusion - Frequently Asked Questions." *IAEA*, United Nations, 12 Oct. 2016, www.iaea.org/topics/energy/fusion/faqs.

16   Edwards, Lin. "The World Is Running out of Helium: Nobel Prize Winner." *Phys.org*, 24 Aug. 2010, phys.org/news/2010-08-world-helium-nobel-prize-winner.html.

17   "Tritium Fact Sheet." *Health Physics Society (HPS)*, Jan. 2020, hps.org/documents/tritium_fact_sheet.pdf.

18   "Tritium: a Challenging Fuel for Fusion." *EUROfusion*, 11 Aug. 2017, www.euro-fusion.org/news/2017-3/tritium-a-challenging-fuel-for-fusion/.

19   "Where Is Coal Found?" *World Coal Association*, 10 Oct. 2019, www.worldcoal.org/coal/where-coal-found.

20   "Tritium (Hydrogen-3)." *Health Physics Society (HPS)*, Oct. 2001, hpschapters.org/northcarolina/NSDS/tritium.pdf.

21   Fetter, Steve. "How Long Will the World's Uranium Supplies Last?" *Scientific American*, 26 Jan. 2009, www.scientificamerican.com/article/how-long-will-global-uranium-deposits-last/#:~:text=If%20the%20Nuclear%20Energy%20Agency%20%28NEA%29%20has%20ac-curately,than%20200%20years%20at%20current%20rates%20of%20consumption.

22   "Nuclear Fusion Basics." *IAEA*, United Nations, 8 Oct. 2010, www.iaea.org/newscenter/news/nuclear-fusion-basics.

23   "Return to the Moon." *Curiosity Stream*, 2018, curiositystream.com/video/2107/return-to-the-moon. There are a plethora of other articles and papers that explore this topic as well.

24   Kramer, David. "ITER Disputes DOE's Cost Estimate of Fusion Project." *Physics Today*, American Institute of Physics, 16 Apr. 2018, physicstoday.scitation.org/do/10.1063/PT.6.2.20180416a/full/.

25 "Each Country's Share of CO2 Emissions." *Union of Concerned Scientists*, 10 Oct. 2019, www.ucsusa.org/resources/each-countrys-share-CO2-emissions.

26 Ibid

27 *World Population Prospects 2019*. 2019, Page 6, *United Nations*, population.un.org/wpp/Publications/Files/WPP2019_Highlights.pdf.

28 Gaughan, Richard. "How Much Land Is Needed for Wind Turbines?" *Sciencing*, 10 May 2018, sciencing.com/much-land-needed-wind-turbines-12304634.html.

29 "Facts & Figures." *ITER*, www.iter.org/FactsFigures.

Chapter 10: The Solution We Need

1   "A Technology Roadmap for Generation IV Nuclear Energy Systems." *U.S. Department of Energy (DOE) Nuclear Energy Research Advisory Committee and the Generation IV International Forum*, Dec. 2002, www.gen-4.org/gif/upload/docs/application/pdf/2013-09/genivroadmap2002.pdf.

2   "Hanford History." *Hanford*, U.S. Department of Energy (DOE), 21 Mar. 21AD, www.hanford.gov/page.cfm/HanfordHistory.

3   Grundhauser, Eric. "The Mysterious Case of the Radioactive Toothpaste." *Atlas Obscura*, 5 May 2017, www.atlasobscura.com/articles/thorium-toothpaste-alsos-world-war-wwii-manhattan-project.

4   Ropeik, David. "The Rise of Nuclear Fear-How We Learned to Fear the Radiation." *Scientific American*, 15 June 2012, blogs.scientificamerican.com/guest-blog/the-rise-of-nuclear-fear-how-we-learned-to-fear-the-bomb/#:~:text=In%20the%20decade%20since%20the%20atomic%20bombs%20were,genie%20of%20nuclear%20fear%20back%20in%20the%20bottle.

5   Ibid

6   "Advanced Nuclear Power Reactors." *World Nuclear Association*, Sept. 2020, world-nuclear.org/information-library/nuclear-fuel-cycle/nuclear-power-reactors/advanced-nuclear-power-reactors.aspx.

7   "U.S. Nuclear Industry." *U.S. Energy Information Administration (EIA)*, 15 Apr. 2020, www.eia.gov/energyexplained/nuclear/us-nuclear-industry.php#:~:text=Although%20some%20foreign%20nuclear%20power%20plants%20have%20as,South%20Carolina.%20All%20three%20plants%20have%20three%20reactors.

8   Chandler, David L. "Explaining the Plummeting Cost of Solar Power." *MIT News*, Massachusetts Institute of Technology (MIT), 20 Nov. 2020, news.mit.edu/2018/explaining-dropping-solar-cost-1120.

9   "USS Nautilus Commissioned." *History.com*, A&E Television Networks, 9 Feb. 2010, www.history.com/this-day-in-history/uss-nautilus-commissioned.

10   Wilson, Taylor. "My Radical Plan for Small Nuclear Fission Reactors." *TED*, Feb. 2013, www.ted.com/talks/taylor_wilson_my_radical_plan_for_small_nuclear_fission_reactors.

11   "Advanced Small Modular Reactors (SMRs)." *U.S. Department of Energy (DOE)*, www.energy.gov/ne/nuclear-reactor-technologies/small-modular-nuclear-reactors.

12   "Small Nuclear Power Reactors." *World Nuclear Association*, Feb. 2021, www.world-nuclear.org/information-library/nuclear-fuel-cycle/nuclear-power-reactors/small-nuclear-power-reactors.aspx.

13   "Time Warp: Molten Salt Reactor Experiment—Alvin Weinberg's Magnum Opus." *Oak Ridge National Laboratory*, U.S. Department of Energy (DOE), www.ornl.gov/molten-salt-reactor/history.

14   Rosenthal, Murray. "MSRE into MSBR?" *Oak Ridge National Laboratory*, U.S. Department of Energy, 1968, www.ornl.gov/sites/default/files/ORNL%20Review%20v2n2%201968.pdf. (Page 1) Emphasis not in original text.

15   Ibid (Page 13)

16   Murray Rosenthal OH final.pdf (oakridgetn.gov) (Page 7)

17   "Uranium Miners (1) - Worker Health Study Summaries." *Centers for Disease Control and Prevention (CDC)*, 11 Apr. 2017, www.cdc.gov/niosh/pgms/worknotify/uranium.html.

18   "Uranium Enrichment." *World Nuclear Association*, Jan. 2020, www.world-nuclear.org/information-library/nuclear-fuel-cycle/conversion-enrichment-and-fabrication/uranium-enrichment.aspx.

19   "Why Highly Enriched Uranium Is a Threat." *Nuclear Threat Initiative*, 1 Aug. 2011, www.nti.org/analysis/articles/why-highly-enriched-uranium-threat/.

20   "Thorium Fueled Reactor Types." *Thorium Energy Alliance*, thoriumenergyalliance.com/5-1-4-thorium-fueled-reactor-types/#:~:text=The%20MSRE%20Never%20used%20Thorium%20as%20a%20Fuel.,fuel%20load%20was%20never%20achieved%20because%20of%20defunding.

21   Alderson, Ella. "The Thorium Fuel Experiments." *Medium*, Predict, 7 Mar. 2020, medium.com/predict/the-thorium-fuel-experiments-40667db70971.

22   Emsley, John. "Thorium." *RSC Education*, 31 Aug. 2010, edu.rsc.org/elements/thorium/2020027.article.

23   "Uranium." *Berkley Rausser*, University of California, nature.berkeley.edu/classes/eps2/wisc/u.html#:~:text=The%20overall%20abundance%20of%20uranium%20in%20the%20Earth's,the%20Soviet%20Union,%20and%20the%20southwestern%20United%20States.

24   "Uranium 233." *Nuclear Power*, www.nuclear-power.net/nuclear-power-plant/nuclear-fuel/uranium/uranium-233/.

25   "Physical, Nuclear, and Chemical Properties of Plutonium." *Institute for Energy and Environmental Research (IEER)*, Apr. 2021, ieer.org/resource/factsheets/plutonium-factsheet/.

26   Moir, Ralph W., "The fusion breeder", Journal of Fusion Energy, Fusion breeder, fusion/fission, hybrid, fusion/fission fuel factory, 1982, oct, 2, 4-5, Pages 351-367, doi = 10.1007/BF01063686, https://ui.adsabs.harvard.edu/abs/1982JFuE....2..351M, Provided by the SAO/NASA Astrophysics Data System

27   "Molten Salt Reactors." *World Nuclear Association*, Dec. 2018, www.world-nuclear.org/information-library/current-and-future-generation/molten-salt-reactors.aspx.

28   "Breeder Reactors." *Science Direct*, Elsevier, www.sciencedirect.com/topics/engineering/breeder-reactors.

29   Gent, Edd. "Why India Wants to Turn Its Beaches into Nuclear Fuel." *BBC Future*, BBC, 18 Oct. 2018, www.bbc.com/future/article/20181016-why-india-wants-to-turn-its-beaches-into-nuclear-fuel.

30   *Thorium, The Far Side of Nuclear (2020)*: 1:01 in . There are other estimates available online as well.

31   Thompson, Elizabeth. "Moon's Largest Crater Holds Clues about Early Lunar Mantle." *Phys.org*, 16 Feb. 2021, phys.org/news/2021-02-moon-largest-crater-clues-early.html.

32   Ma, Chanthia. "Mythbusters: Is There Really a Dark Side of the Moon?" *Yale Scientific Magazine*, 8 May 2015, www.yalescientific.org/2015/05/mythbusters-is-there-really-a-dark-side-of-the-moon/#:~:text=In%20fact%2C%20the%20far%20side%20of%20the%20moon,except%20in%20the%20case%20of%20a%20lunar%20eclipse.

33   Pedersen, Thomas J., speaker. *Making Safe Nuclear Power from Thorium*. *YouTube*, TEDx Talks, 15 Nov.

2016,www.youtube.com/watch?v=tHO1ebNxhVI&list=PLI0ToICTHWJGvDTEs8rjNstI2m_
p7IXxl&index=6&t=318s.

34   "A Technology Roadmap for Generation IV Nuclear Energy Systems." *U.S. Department of Energy (DOE) Nuclear Energy Research Advisory Committee and the Generation IV International Forum*, Dec. 2002, www.gen-4.org/gif/upload/docs/application/pdf/2013-09/genivroadmap2002.pdf.

35   "Advanced Reactor Demonstration Program." *U.S. Department of Energy (DOE)*, www.energy.gov/ne/nuclear-reactor-technologies/advanced-reactor-demonstration-program.

36   IEA (2020), Nuclear Power, IEA, Paris https://www.iea.org/reports/nuclear-power

37   Wald, Matt. "Andrew Yang Loves Thorium Nuclear Reactors, But What Are They?" *Nuclear Energy Institute*, 12 Sept. 2019, www.nei.org/news/2019/andrew-yang-loves-thorium-reactors.

38   "Plan for Climate Change and Environmental Justice: Joe Biden." *Joe Biden for President: Official Campaign Website*, joebiden.com/climate-plan/.

Chapter 11: Risking Destruction

1   "Nazi Laws." *Department of Financial Services*, New York State, www.dfs.ny.gov/system/files/documents/2020/08/nazi_laws_summary_english.pdf.

2   "Einstein and the Manhattan Project." *American Museum of Natural History*, www.amnh.org/exhibitions/einstein/peace-and-war/the-manhattan-project.

3   Horgan, John. "Bethe, Teller, Trinity and the End of Earth." *Scientific American*, 4 Aug. 2015, blogs.scientificamerican.com/cross-check/bethe-teller-trinity-and-the-end-of-earth/.

4   Achenbach, Joel. "The Man Who Feared, Rationally, That He'd Just Destroyed the World." *The Washington Post*, 7 Apr. 2019, www.washingtonpost.com/news/achenblog/wp/2015/07/23/the-man-who-feared-rationally-that-hed-just-destroyed-the-world/.

5   "Nuclear Share of Electricity Generation in 2020." *Power Reactor Information System (PRIS)*, International Atomic Energy Agency (IAEA), pris.iaea.org/PRIS/WorldStatistics/Nuclear-ShareofElectricityGeneration.aspx. (Pulled from charts) These charts include Taiwan (who operates 4 nuclear reactors) with mainland China, though a distinction is made at the bottom. There are 30 countries counted, but considering Taiwan as its own country (as I do in the text) gives you 33 countries total.

6   Ibid

7   Ibid

8   "Megatons to Megawatts." *Centrus Energy Corp*, www.centrusenergy.com/who-we-are/history/megatons-to-megawatts/.

9   "Countries With Nuclear Weapons 2020." *World Population Review*, worldpopulationreview.com/countries/countries-with-nuclear-weapons/.

10   The documentary *Zero-days* does a great job at exploring the implications of the Stuxnet virus attack.

11   "Proliferation Resistance & Physical Protection Working Group (PRPPWG)." *Gen. IV International Forum*, www.gen-4.org/gif/jcms/c_40411/proliferation-resistance-physical-protection-working-group-prppwg.

12   "Proliferation Resistance and Physical Protection of the Six Generation IV Nuclear Energy Systems." *The Proliferation Resistance and Physical Protection Evaluation Methodology Working Group and the System Steering Committees of the Generation IV International Forum*, Gen. IV International Forum, 15 July 2015, www.gen-4.org/gif/upload/docs/application/pdf/2013-09/gif_prppwg_ssc_report_final.pdf. (Pages 108-113)

13   Ibid (Page 111)

14   A. Glaser and R.J. Goldston 2012 Nucl. Fusion 52 043004, "Proliferation risks of magnetic fusion energy: clandestine production, covert production and breakout", DOI: https://doi.org/10.1088/0029-5515/52/4/043004

15   Ibid

16   "The Costs of the Manhattan Project.", *Brookings*, 14 Apr. 2017, www.brookings.edu/the-costs-of-the-manhattan-project/.

17   Ibid

18   Vaez, Ali, and Karim Sadjadpour. "Iran's Nuclear Odyssey: Costs and Risks." *Carnegie Endowment for International Peace*, Federation of American Scientists, 2013, fas.org/wp-content/uploads/2013/04/iran_nuclear_odyssey.pdf.

19   "Six Charts That Show How Hard US Sanctions Have Hit Iran." *BBC News*, 9 Dec. 2019, www.bbc.com/news/world-middle-east-48119109.

20   "Islamic Republic of Iran- Overview." *The World Bank*, www.worldbank.org/en/country/iran/overview.

21   Stetka, Bret. "Steven Pinker: This Is History's Most Peaceful Time—New Study: 'Not So Fast.'" *Scientific American*, 9 Nov. 2017, www.scientificamerican.com/article/steven-pinker-this-is-historys-most-peaceful-time-new-study-not-so-fast/

22   Roser, Max. "War and Peace." *Our World in Data*, 13 Dec. 2016, ourworldindata.org/war-and-peace#war-and-peace-after-1945. (I highly recommend looking through these charts as they put the current state of the world into a relaxing context, as well as Steven Pinker's book, *Enlightenment Now*.)

Chapter 12: The Problems We Ignore

1   "Fast Facts on Transportation Greenhouse Gas Emissions." *U.S. Environmental Protection Agency (EPA)*, 29 July 2020, www.epa.gov/greenvehicles/fast-facts-transportation-greenhouse-gas-emissions.

2   Palmer, Annie. "Amazon Debuts Electric Delivery Vans Created with Rivian." *CNBC*, 8 Oct. 2020, www.cnbc.com/2020/10/08/amazon-new-electric-delivery-vans-created-with-rivian-unveiled.html.

3   "Sectoral Greenhouse Gas Emissions by IPCC Sector." *European Environment Agency (EEA)*, 21 June 2016, www.eea.europa.eu/data-and-maps/daviz/change-of-co2-eq-emissions-2#tab-dashboard-01.

4   Matulka, Rebecca. "#AskEnergySaver: Home Energy Audits." *U.S. Department of Energy (DOE)*, 24 Jan. 2014, www.energy.gov/articles/askenergysaver-home-energy-audits. Infographic by <a href="/node/379579">Sarah Gerrity</a>, Energy Department

5   "Apple Commits to Be 100 Percent Carbon Neutral for Its Supply Chain and Products by 2030." *Apple Newsroom*, 21 July 2020, www.apple.com/newsroom/2020/07/apple-commits-to-be-100-percent-carbon-neutral-for-its-supply-chain-and-products-by-2030/.

6   "Emissions and Climate." *ExxonMobil*, corporate.exxonmobil.com/Sustainability/Emissions-and-climate.

Chapter 13: Moving the World

1   "Natural Gas Composition." *Croft Productions Systems, Inc.*, 01 May 2020, www.croftsystems.net/oil-gas-blog/natural-gas-composition/.

2   Ge, Mengpin, and Johannes Friedrich. "4 Charts Explain Greenhouse Gas Emissions by Countries and Sectors." *World Resources Institute (WRI)*, 6 Feb. 2020, www.wri.org/blog/2020/02/greenhouse-gas-emissions-by-country-sector. (Pulled from graph "World Greenhouse Gas Emissions in 2016 (Sector | End Use | Gas)")

3   "U.S. Vehicle-Miles." *Bureau of Transportation Statistics*, U.S. Department of Transportation, www.bts.gov/content/us-vehicle-miles. (Pulled from table "U.S. Vehicle-Miles")

4   "Fast Facts on Transportation Greenhouse Gas Emissions." *U.S. Environmental Protection Agency (EPA)*, 29 July 2020, www.epa.gov/greenvehicles/fast-facts-transportation-greenhouse-gas-emissions.

5   "VEHICLE TRAVEL BY SELECTED COUNTRY (Metric)." *Federal Highway Administration (FHWA)*, U.S. Department of Transportation, Apr. 2010, www.fhwa.dot.gov/policyinformation/statistics/2008/pdf/in5.pdf.

6   "Who Built the First Automobile?" *History.com*, A&E Television Networks, 11 Dec. 2012, www.history.com/news/who-built-the-first-automobile.

7   "1896 Riker Electric Tricycle." *The Henry Ford Organization*, www.thehenryford.org/collections-and-research/digital-collections/artifact/274188/.

8   "Electric Fire Engine." *Wikipedia*, Wikimedia Foundation, 7 Dec. 2020, en.wikipedia.org/wiki/Electric_fire_engine#CITEREFGovernment_Printing_Office1886.

9   Alarco, Jose, and Peter Talbot. "The History and Development of Batteries." *Phys.org*, 30 Apr. 2015, phys.org/news/2015-04-history-batteries.html.

10   Stevens, Pippa. "The Battery Decade: How Energy Storage Could Revolutionize Industries in the next 10 Years." *CNBC*, 30 Dec. 2019, www.cnbc.com/2019/12/30/battery-developments-in-the-last-decade-created-a-seismic-shift-that-will-play-out-in-the-next-10-years.html.

11   "Lithium-Ion Battery." *Clean Energy Institute*, University of Washington, www.cei.washington.edu/education/science-of-solar/battery-technology/.

12   Katz-Lavigne, Sarah. "Demand for Congo's Cobalt Is on the Rise. So Is the Scrutiny of Mining Practices." *The Washington Post*, WP Company, 21 Feb. 2019, www.washingtonpost.com/politics/2019/02/21/demand-congos-cobalt-is-rise-so-is-scrutiny-mining-practices/.

13   Scott, Joe. "The Trouble With Cobalt." *Answers With Joe*, YouTube, 24 Feb. 2020, www.youtube.com/watch?v=25G4BcioPjE.

14   IEA (2011), Energy Efficiency Policy Opportunities for Electric Motor-Driven Systems, IEA, Paris https://www.iea.org/reports/energy-efficiency-policy-opportunities-for-electric-motor-driven-systems

15   Roof, Katie, and Edward Ludlow. "Bloomberg.com." *Bloomberg*, 5 Jan. 2021, www.bloomberg.com/news/articles/2021-01-05/rivian-is-said-close-to-raising-funds-at-25-billion-valuation.

16   Kolodny, Lora, and Michael Wayland. "Tesla's Market Cap Tops the 9 Largest Automakers Combined - Experts Disagree about If That Can Last." *CNBC*, 14 Dec. 2020, www.cnbc.com/2020/12/14/tesla-valuation-more-than-nine-largest-carmakers-combined-why.html.

17   "Electric Vehicle Sales: Facts & Figures." *Edison Electric Institute (EEI)*, Apr. 2019, www.eei.org/issuesandpolicy/electrictransportation/documents/final_ev_sales_update_april20 19.pdf.

18   Plumer, Brad, and Jill Cowan. "California Plans to Ban Sales of New Gas-Powered Cars in 15 Years." *The New York Times*, 23 Sept. 2020, www.nytimes.com/2020/09/23/climate/california-ban-gas-cars.html.

19   Owuor, Sophy. "Countries That Will Ban Gasoline Cars." *WorldAtlas*, 19 Dec. 2018, www.worldatlas.com/articles/countries-that-will-ban-gasoline-cars.html#:~:text=Countries%20That%20Will%20Ban%20Gasoline%20Cars%20%20,%20%202040%20%2013%20more%20rows%20.

20   Tirupati, Manoz Kumar M. "An Overview on Electric Vehicle Charging Infrastructure." *Tata Elxsi*, 2019, www.tataelxsi.com/Perspectives/WhitePapers/TataElxsi_Whitepaper_An_Overview_on_Electric_Vehicle_Charging_Infrastructure.pdf. (Page 12)

21   "List of Fuel Cell Vehicles." *Wikipedia*, Wikimedia Foundation, 26 Feb. 2021, en.wikipedia.org/wiki/List_of_fuel_cell_vehicles.

22   "Advanced Biofuels." *ExxonMobil*, corporate.exxonmobil.com/Energy-and-innovation/Advanced-biofuels.

23   "Natural Gas Vehicles." *Alternative Fuels Data Center*, U.S. Department of Energy (DOE), https://afdc.energy.gov/vehicles/natural_gas.html#:~:text=Natural%20gas%20powers%20more%20than%20175%2C000%20vehiclesin%20the,long-haul%20routes%20where%20fuel%20stations%20can%20become%20sparse

24   "Natural Gas Benefits and Considerations." *Alternative Fuels Data Center*, U.S. Department of Energy (DOE), afdc.energy.gov/fuels/natural_gas_benefits.html.

25   Copy William Harris "How Natural-gas Vehicles Work" 9 September 2005. HowStuffWorks.com., https://auto.howstuffworks.com/fuel-efficiency/alternative-fuels/ngv.htm,

26   IEA (2020), Global EV Outlook 2020, IEA, Paris https://www.iea.org/reports/global-ev-outlook-2020 (Pulled from graph)

27   Brown, Alex. "Electric Cars Will Challenge State Power Grids." *The Pew Charitable Trusts*, 9 Jan. 2020, www.pewtrusts.org/en/research-and-analysis/blogs/stateline/2020/01/09/electric-cars-will-challenge-state-power-grids.

28   McLaren, Joyce, et al. "Emissions Associated with Electric Vehicle Charging: Impact of Electricity Generation Mix, Charging Infrastructure Availability, and Vehicle Type." *National Renewable Energy Laboratory (NREL)*, U.S. Department of Energy, Apr. 2016, afdc.energy.gov/files/u/publication/ev_emissions_impact.pdf. NREL/TP-6A20-64852

29   Nealer, Rachael, et al. "Cleaner Cars from Cradle to Grave: How Electric Cars Beat Gasoline Cars on Lifetime Global Warming Emissions." *Union of Concerned Scientists*, Nov. 2015, www.ucsusa.org/sites/default/files/attach/2015/11/Cleaner-Cars-from-Cradle-to-Grave-full-report.pdf.

30   Hong Huo, Qiang Zhang, Michael Q. Wang, David G. Streets, and Kebin He *Environmental Science & Technology* 2010 *44* (13), 4856-4861 DOI: 10.1021/es100520c

31   "Greenhouse Gas Emissions from Electric and Plug-In Hybrid Vehicles." *Office of Energy Efficiency and Renewable Energy*, U.S. Department of Energy (DOE), www.fueleconomy.gov/feg/Find.do?action=bt2.

32   Kundu, A., Shul, Y. G., & Kim, D. H. (2007). Chapter Seven Methanol Reforming Processes. In *Advances in Fuel Cells* (pp. 419-472). (Advances in Fuel Cells; Vol. 1). Elsevier Ltd. https://doi.org/10.1016/S1752-301X(07)80012-3

33   "Hydrogen Fuel Basics." *U.S. Department of Energy (DOE)*, www.energy.gov/eere/fuelcells/hydrogen-fuel-basics.

34   Josh Clark "Is hydrogen fuel dangerous?" 19 May 2008, HowStuffWorks.com https://auto.howstuffworks.com/fuel-efficiency/alternative-fuels/dangerous-hydrogen-fuel.htm

35   IEA (2020), Aviation, IEA, Paris https://www.iea.org/reports/aviation

36   IEA, Transport sector CO2 emissions by mode in the Sustainable Development Scenario, 2000-2030, IEA, Paris https://www.iea.org/data-and-statistics/charts/transport-sector-co2-emissions-by-mode-in-the-sustainable-development-scenario-2000-2030

37   IEA, Atlas of Energy, IEA, Paris http://energyatlas.iea.org/#!/tellmap/1378539487

38   Nilsen, Thomas. "Electric Aviation Ready for Take-off in Norway by 2030, Report Says." *The Barents Observer*, 8 Mar. 2020, thebarentsobserver.com/en/travel/2020/03/electric-aviation-should-be-ready-take-norway-2030.

39   "Fast Facts: U.S. Transportation Sector Greenhous Gas Emissions 1990-2018." *U.S. Environmental Protection Agency (EPA)*, June 2020, nepis.epa.gov/Exe/ZyPDF.cgi?Dockey=P100 ZK4P.pdf. (Pulled from chart "U.S. Transportation GHG Emissions" on page 2)

40   Ibid

41   "Third IMO GHG Study 2014." *International Maritime Organization (IMO)*, 2014, www.imo.org/en/OurWork/Environment/Pages/Greenhouse-Gas-Studies-2014.aspx. (Point 1.1)

42   Ibid (Point 5.1)

43   "Integrating Maritime Transport Emissions in the EU's Greenhouse Gas Reduction Policies." *European Commission*, 2013, ec.europa.eu/clima/sites/clima/files/transport/shipping/docs/com_2013_479_en.pdf.

44   IEA (2020), Rail, IEA, Paris https://www.iea.org/reports/rail

45   Ibid

46   Ritchie, Hannah. "Cars, Planes, Trains: Where Do CO2 Emissions from Transport Come from?" *Our World in Data*, 6 Oct. 2020, ourworldindata.org/co2-emissions-from-transport.

47   *California High-Speed Rail Authority*, hsr.ca.gov/programs/private_property/.

48   Vartabedian, Ralph. "California's Troubled Bullet Train Project Getting One of Biggest Management Upheavals in Years." *Los Angeles Times*, 26 July 2019, www.latimes.com/california/story/2019-07-26/california-high-speed-rail-authority-management-shakeup.

49   IEA (2020), Rail, IEA, Paris https://www.iea.org/reports/rail

50   Huddleston Jr., Tom. "This Is What Traveling on Elon Musk's 700 Mph Hyperloop Could Look Like." *CNBC*, 17 July 2019, www.cnbc.com/2019/07/17/images-what-elon-musks-hyperloop-could-look-like.html#:~:text=Elon%20Musk%20is%20not%20the%20only%20billionaire%20working,by%20the%20%E2%80%9Cmid-2020s%E2%80%9D,%20the%20company%20said%20in%202018.

51   IEA (2020), Energy Technology Perspectives 2020, IEA, Paris https://www.iea.org/reports/energy-technology-perspectives-2020 (Pulled from graph "CO2 emissions reductions in the [transport] sector in the Sustainable Development Scenario relative to the Stated Policies Scenario")

52 "Each Country's Share of CO2 Emissions." *Union of Concerned Scientists*, 12 Aug. 2020, www.ucsusa.org/resources/each-countrys-share-co2-emissions. (Pulled from table "The 20 countries that emitted the most carbon dioxide in 2018")

Chapter 14: Making the World

1   "Sources of Greenhouse Gas Emissions." *U.S. Environmental Protection Agency (EPA)*, 4 Dec. 2020, www.epa.gov/ghgemissions/sources-greenhouse-gas-emissions.

2   "Each Country's Share of CO2 Emissions." *Union of Concerned Scientists*, 12 Aug. 2020, www.ucsusa.org/resources/each-countrys-share-co2-emissions. (Pulled from table "The 20 countries that emitted the most carbon dioxide in 2018")

3   IEA (2020), Tracking Industry 2020, IEA, Paris https://www.iea.org/reports/tracking-industry-2020

4   IEA (2020), Tracking Industry 2020, IEA, Paris https://www.iea.org/reports/tracking-indus-try-2020 (Pulled from graph "Industry direct CO2 emissions in the Sustainable Development Scenario, 2000-2030"))

5   Ibid

6   IEA (2020), Cement, IEA, Paris https://www.iea.org/reports/cement

7   Timperley, Jocelyn. "Q&A: Why Cement Emissions Matter for Climate Change." *Carbon Brief*, 13 Sept. 2018, www.carbonbrief.org/qa-why-cement-emissions-matter-for-climate-change.

8   Olivier, Jos G.J., et al "Trends in Global CO2 Emissions: 2016 Report." *Netherlands Environmental Assessment Agency & Joint Research Centre*, European Commission, 2016, edgar.jrc.ec.europa.eu/news_docs/jrc-2016-trends-in-global-co2-emissions-2016-report-103425.pdf. Publication number: 2315, JRC Science for Policy Report: 103428 (Pages 64-66)

9   IEA (2020), Cement, IEA, Paris https://www.iea.org/reports/cement

10   Ibid

11   Schuler, Tom. "Transcript of 'How We Could Make Carbon-Negative Concrete.'" *TED*, 4 Jan. 2021, www.ted.com/talks/tom_schuler_how_we_could_make_carbon_negative_concrete/transcript#t-277254.

12   IEA (2020), Cement, IEA, Paris https://www.iea.org/reports/cement

13   *The World In a Grain*, by Vince Beiser, does a fantastic job of exploring this topic in depth.

14   Edwards, Bruce. "The Insatiable Demand for Sand." *International Monetary Fund (IMF)*, Dec. 2015, www.imf.org/external/pubs/ft/fandd/2015/12/edwards.htm.

15   "Sand, Rarer than One Thinks." *United Nations Environment Programme (UNEP)*, Mar. 2014, na.unep.net/geas/getUNEPPageWithArticleIDScript.php?article_id=110.

16   Ludacer, Rob. "The World Is Running out of Sand - and There's a Black Market for It Now." *Business Insider*, 11 June 2018, www.businessinsider.com/world-running-out-sand-resources-concrete-2018-6.

17   "Sand, Rarer than One Thinks." *United Nations Environment Programme (UNEP)*, Mar. 2014, na.unep.net/geas/getUNEPPageWithArticleIDScript.php?article_id=110. (See "Box 1: Dubai")

18   Ibid (See "Box 2: Singapore")

19   Ibid

20   Pearce, Fred. "The Hidden Environmental Toll of Mining the World's Sand." *Yale Environment 360*, Yale School of the Environment, 5 Feb. 2019, e360.yale.edu/features/the-hid-den-environmental-toll-of-mining-the-worlds-sand.

21   IEA (2020), Cement, IEA, Paris https://www.iea.org/reports/cement

22   IEA (2020), Iron and Steel, IEA, Paris https://www.iea.org/reports/iron-and-steel (Pulled from graph "Energy demand and intensity in iron and steel, 2000-2018")

23   "Fossil Free Steel. Another Giant Step towards Net Carbon Zero?" *Just Have a Think*, YouTube, 13 Dec. 2020, www.youtube.com/watch?v=ywHJt88H5YQ.

24   IEA (2020), Iron and Steel, IEA, Paris https://www.iea.org/reports/iron-and-steel (Pulled from graph "Energy demand and intensity in iron and steel, 2000-2018")

25   "Global Crude Steel Output Increases by 3.4% in 2019." *Worldsteel Association* , 27 Jan. 2020, www.worldsteel.org/media-centre/press-releases/2020/Global-crude-steel-output-increases-by-3.4—in-2019.html.

26   IEA (2020), Iron and Steel, IEA, Paris https://www.iea.org/reports/iron-and-steel

27   Ibid

28   IEA (2020), Chemicals, IEA, Paris https://www.iea.org/reports/chemicals

29   IEA (2020), CCUS in Industry and Transformation, IEA, Paris https://www.iea.org/reports/ccus-in-industry-and-transformation

Chapter 15: Living in the World

1   This figure includes not just GHG emissions directly coming from buildings but also their energy requirements for heating and electricity. If heating and electricity are included outside of the "buildings" sector (as it is in most references in this book) then buildings account for just 6.5% of anthropogenic GHG emissions.
IPCC, 2014: Climate Change 2014: Synthesis Report. Contribution of Working Groups I, II and III to the Fifth Assessment Report of the Intergovernmental Panel on Climate Change [Core Writing Team, R.K. Pachauri and L.A. Meyer (eds.)]. IPCC, Geneva, Switzerland, (Page 47, pulled from figure 1.7)

2   IEA (2020), Tracking Buildings 2020, IEA, Paris https://www.iea.org/reports/tracking-buildings-2020

3   Ibid (Pulled from graph "Building sector energy-related CO2 emissions in the Sustainable Development Scenario, 2000-2030")

4   IEA (2020), Appliances and Equipment, IEA, Paris https://www.iea.org/reports/appliances-and-equipment (Pulled from graph "Key drivers of changes in household appliance energy consumption in the Sustainable Development Scenario, 2000-2030")

5   Holst, Arne. "Global AC Penetration Rate by Country 2016." *Statista*, 2 Mar. 2020, www.statista.com/statistics/911064/worldwide-air-conditioning-penetration-rate-country/.

6   IEA (2020), Cooling, IEA, Paris https://www.iea.org/reports/cooling (Pulled from graph "Global space cooling equipment sales in 2019 and compound annual growth rate, 2010 19")

7   Ibid

8   "Recent International Developments under the Montreal Protocol." *U.S. Environmental Protection Agency (EPA)*, 26 Sept. 2018, www.epa.gov/ozone-layer-protection/recent-international-developments-under-montreal-protocol.

Chapter 16: Helping Mother Nature

1   Harvey, Chelsea. "CO2 Emissions Reached an All-Time High in 2018." *Scientific American*, E&E News, 6 Dec. 2018, www.scientificamerican.com/article/CO2-emissions-reached-an-all-time-high-in-2018/.

2   IEA (2021), Global Energy Review: CO2 Emissions in 2020, IEA, Paris https://www.iea.org/articles/global-energy-review-co2-emissions-in-2020 (Pulled from graph "Global energy-related CO2 emissions, 1990-2020")

3   IPCC, 2014: Climate Change 2014: Synthesis Report. Contribution of Working Groups I, II and III to the Fifth Assessment Report of the Intergovernmental Panel on Climate Change [Core Writing Team, R.K. Pachauri and L.A. Meyer (eds.)]. IPCC, Geneva, Switzerland (Page 45)

4   Ibid (Page 47)

5   "Global Ocean Is Absorbing More Carbon from Fossil Fuel Emissions." *National Oceanic and Atmospheric Administration (NOAA)*, www.noaa.gov/news/global-ocean-is-absorbing-more-carbon-from-fossil-fuel-emissions.

6   Ibid

7   "Global Warming of 1.5 °C Special Report." *Intergovernmental Panel on Climate Change (IPCC)*, www.ipcc.ch/sr15/.

8   Taylor, Ashley P. "Earth Has Room For 1 Trillion More Trees." *The Scientist*, 5 July 2019, www.the-scientist.com/news-opinion/earth-has-room-for-1-trillion-more-trees-66103.

9   Borenstein, Seth. "Best Way to Fight Climate Change? Plant a Trillion Trees." *AP NEWS*, Associated Press, 4 July 2019, apnews.com/8ac33686b64a4fbc991997a72683b1c5.

10   Bastin, Jean-Francois, *et al.* "The Global Tree Restoration Potential." *Science Journal*, American Association for the Advancement of Science, 2019, science.sciencemag.org/content/365/6448/76.abstract. DOI: 10.1126/science.aax0848

11   Griscom, Bronson W., et al. "Natural Climate Solutions." *Proceedings of the National Academy of Sciences*, vol. 114, no. 44, 2017, pp. 11645–11650., doi:10.1073/pnas.1710465114.

12   Popkin, Gabriel. "How Much Can Forests Fight Climate Change?" *Nature News*, Nature Publishing Group, 15 Jan. 2019, www.nature.com/articles/d41586-019-00122-z. *Nature* 565, 280-282 (2019) doi: https://doi.org/10.1038/d41586-019-00122-z

13   McLaren, Duncan. "Exaggerating How Much CO_2 Can Be Absorbed by Tree Planting Risks Deterring Crucial Climate Action." *The Conversation*, 12 July 2019, theconversation.com/exaggerating-how-much-co-can-be-absorbed-by-tree-planting-risks-deterring-crucial-climate-action-120170.

14   Crowther, T., Glick, H., Covey, K. *et al.* Mapping tree density at a global scale. *Nature* 525, 201–205 (2015). https://doi.org/10.1038/nature14967

15   Song, XP., Hansen, M.C., Stehman, S.V. *et al.* Global land change from 1982 to 2016. *Nature* 560, 639–643 (2018). https://doi.org/10.1038/s41586-018-0411-9

16   DeMartini, Alayna. "Higher Carbon Dioxide Levels Prompt More Plant Growth, But Fewer Nutrients." *College of Food, Agricultural, and Environmental Sciences (CFAES)*, 3 Apr. 2018, cfaes.osu.edu/news/articles/higher-carbon-dioxide-levels-prompt-more-plant-growth-fewer-nutrients.

17   Kirk, Karin. "More CO2 in the Atmosphere Hurts Key Plants and Crops More than It Helps." *Yale Climate Connections*, 13 Dec. 2020, yaleclimateconnections.org/2020/12/more-co2-in-the-atmosphere-hurts-key-plants-and-crops-more-than-it-helps/.
The following is an example of one such study conducted by the EPA. It becomes apparent quickly into the study report that this is a costly and time-consuming initiative, something that could greatly hinder widespread tree planting in some areas.
18   Smith, G.F.1, Gittings, T.2, Wilson, M. *et al.* "Biodiversity Assessment of Afforestation Sites." *U.S. Environmental Protection Agency (EPA)*, www.epa.ie/pubs/reports/research/biodiversity/bioforestfinalreport/2000-LS-311-M2%20Final%20Report-MAY%202006.pdf.
19   "The Great Green Wall." *Great Green Wall*, www.greatgreenwall.org/about-great-green-wall.
20   IEA (2020), Energy Technology Perspectives 2020, IEA, Paris https://www.iea.org/reports/energy-technology-perspectives-2020 (Pulled from graph "CO2 emissions reductions in the [energy, power, industry, heavy industry] sector in the Sustainable Development Scenario relative to the Stated Policies Scenario")
21   IEA (2020), CCUS in Industry and Transformation, IEA, Paris https://www.iea.org/reports/ccus-in-industry-and-transformation (Pulled from map "CCUS projects in industry")
22   IEA (2020), CCUS in Power, IEA, Paris https://www.iea.org/reports/ccus-in-power (Pulled from map)
23   Ibid
24   "Carbon Capture and Storage: a Technological Challenge Already Solved." *UNECE*, United Nations, www.unece.org/unece-and-the-sdgs/climate-change/sustainable-development-climate-changeunece-and-climate-change/carbon-capture-and-storage-a-technological-challenge-already-solved.html.

Chapter 17: The Need for Action

1   "Svante Arrhenius (1859-1927)." *NASA*, 18 Jan. 2000, earthobservatory.nasa.gov/features/Arrhenius/arrhenius_3.php.

2   Fagan, Moira, and Christine Huang. "A Look at How People around the World View Climate Change." *Pew Research Center*, 18 Apr. 2019, www.pewresearch.org/fact-tank/2019/04/18/a-look-at-how-people-around-the-world-view-climate-change/.

3   Buchholz, Katharina. "Infographic: Where Climate Change Deniers Live." *Statista Infographics*, 3 Dec. 2020, www.statista.com/chart/19449/countries-with-biggest-share-of-climate-change-deniers/.

4   "Yale Climate Opinion Maps 2020." *Yale Program on Climate Change Communication*, 2 Sept. 2020, climatecommunication.yale.edu/visualizations-data/ycom-us/.

The following are just a few of these studies. This list was compiled and copied from the next shown source (citation #6 for this chapter).

J. Cook, et al, "Consensus on consensus: a synthesis of consensus estimates on human-caused global warming," *Environmental Research Letters* Vol. 11 No. 4, (13 April 2016); DOI:10.1088/1748-9326/11/4/048002

J. Cook, et al, "Quantifying the consensus on anthropogenic global warming in the scientific literature," *Environmental Research Letters* Vol. 8 No. 2, (15 May 2013); DOI:10.1088/1748-9326/8/2/024024

W. R. L. Anderegg, "Expert Credibility in Climate Change," *Proceedings of the National Academy of Sciences* Vol. 107 No. 27, 12107-12109 (21 June 2010); DOI: 10.1073/pnas.1003187107.

P. T. Doran & M. K. Zimmerman, "Examining the Scientific Consensus on Climate Change," *Eos Transactions American Geophysical Union* Vol. 90 Issue 3 (2009), 22; DOI: 10.1029/2009EO030002.

N. Oreskes, "Beyond the Ivory Tower: The Scientific Consensus on Climate Change," *Science* Vol. 306 no. 5702, p. 1686 (3 December 2004); DOI: 10.1126/science.1103618.

5   "Scientific Consensus: Earth's Climate Is Warming." *NASA*, climate.nasa.gov/scientific-consensus/#footnote_1.

6   *Climate Consensus Letter*. 21 Oct. 2009, www.aaas.org/sites/default/files/1021climate_letter1.pdf.

7   USGCRP, 2018: *Impacts, Risks, and Adaptation in the United States: Fourth National Climate Assessment, Volume II* [Reidmiller, D.R., C.W. Avery, D.R. Easterling, K.E. Kunkel, K.L.M. Lewis, T.K. Maycock, and B.C. Stewart (eds.)]. U.S. Global Change Research Program, Washington, DC, USA, 1515 pp. doi: 10.7930/NCA4.2018.

8   "Agencies." *U.S. Global Change Research Program*, www.globalchange.gov/agencies.

9   Jones, Zoe Christen. "Scientific American Backs Joe Biden for President, Its First Endorsement in 175-Year History." *CBS News*, 15 Sept. 2020, www.cbsnews.com/news/scientific-american-endorses-joe-biden-president/.

10   "List of Worldwide Scientific Organizations." *Governors Office of Planning and Research*, State of California, opr.ca.gov/facts/list-of-scientific-organizations.html.

11   Bast, Diane. "30,000 Scientists Sign Petition on Global Warming." *The Heartland Institute*, 1 July 2008, www.heartland.org/news-opinion/news/30000-scientists-sign-petition-on-global-warming.

12   Grandia, Kevin. "The 30,000 Global Warming Petition Is Easily-Debunked Propaganda." *HuffPost*, 22 Aug. 2009, www.huffpost.com/entry/the-30000-global-warming_b_243092.

13   Trenberth, Kevin. "Brouhaha over Hacked Climate Emails." *CGD's Climate Analysis Section (CAS)*, 2009, web.archive.org/web/20100611224809/www.cgd.ucar.edu/cas/Trenberth/statement.html.

14   Study citation for reference: Trenberth, Kevin E. "An Imperative for Climate Change Planning: Tracking Earth's Global Energy." *Current Opinion in Environmental Sustainability*, vol. 1, no. 1, 2009, pp. 19–27., doi:10.1016/j.cosust.2009.06.001.

15   "Global Warming Skeptic Organizations (2013)." *Union of Concerned Scientists*, 16 Aug. 2013, www.ucsusa.org/resources/global-warming-skeptic-organizations.

16   Both the documentary and the book are simply fantastic. I would personally recommend watching the documentary and then reading the book, as the book goes in much deeper detail.

17   "Volcano Hazards Program." *U.S. Geological Survey*, www.usgs.gov/natural-hazards/volcano-hazards/volcanoes-can-affect-climate#:~:text=Do%20the%20Earth%27s%20volcanoes%20emit%20more%20CO%202,%20%200.22%20%203%20more%20rows%20.

Chapter 18: Habitual Suffering

1   Jerving, Sara. *et al* "What Exxon Knew about the Earth's Melting Arctic." *Los Angeles Times*, 9 Oct. 2015, graphics.latimes.com/exxon-arctic/.

2   Carpenter, Zoë. "Exxon Won a Major Climate Change Lawsuit-but More Are Coming." *The Nation*, 13 Dec. 2019, www.thenation.com/article/archive/exxon-lawsuit-climate-change/.

3   Banerjee, Neela; Song, Lisa and David Hasemyer. "Exxon's Own Research Confirmed Fossil Fuels' Role in Global Warming Decades Ago." *InsideClimate News*, 16 Sept. 2015, insideclimate-news.org/news/15092015/Exxons-own-research-confirmed-fossil-fuels-role-in-global-warming.

4   Hall, Shannon. "Exxon Knew about Climate Change Almost 40 Years Ago." *Scientific American*, 26 Oct. 2015, www.scientificamerican.com/article/exxon-knew-about-climate-change-almost-40-years-ago/.

5   Ibid

6   "What Is U.S. Electricity Generation by Energy Source." *EIA*, United States, 25 Oct. 2019, www.eia.gov/tools/faqs/faq.php?id=427&t=3.

7   "Safety of Nuclear Power Reactors: Appendices." *World Nuclear Association*, Mar. 2017, www.world-nuclear.org/information-library/safety-and-security/safety-of-plants/appendices/safety-of-nuclear-power-reactors-appendix.aspx.

8   Duzgun, H. Sebnem, and Nancy Levenson. *Safety Science,* . Part A ed., vol. 110, *Analysis of Soma Mine Disaster Using Causal Analysis Based on Systems Theory (CAST)*. pg. 37–57, ScienceDirect, 2018, reader.elsevier.com/reader/sd/pii/S0925753516303769?token=4AFBE72643E2989E796D4288614226718FFBB54A6828B820A02969AC31795FCD346B3D491E8629759790EE760232691C.

9   "1998 Jesse Pipeline Explosion." *Wikipedia*, Wikimedia Foundation, 23 Aug. 2017, 01:25, en.wikipedia.org/wiki/1998_Jesse_pipeline_explosion.

10   "China Tries to Plug Burst Natural Gas Well." *NBCNews.com*, NBCUniversal News Group, 25 Dec. 2003, www.nbcnews.com/id/3805629#.XkM8UzFKiUk.

11   "Particulate Matter (PM) Basics." *EPA*, Environmental Protection Agency, 14 Nov. 2018, www.epa.gov/pm-pollution/particulate-matter-pm-basics.

12   "What Is Acid Rain?" *EPA*, Environmental Protection Agency, 20 Dec. 2019, www.epa.gov/acidrain/what-acid-rain.

13   *WHO Air Quality Guidelines for Particulate Matter, Ozone, Nitrogen Dioxide and Sulfur Dioxide.* 2005, *WHO Air Quality Guidelines for Particulate Matter, Ozone, Nitrogen Dioxide and Sulfur Dioxide,* apps.who.int/iris/bitstream/handle/10665/69477/WHO_SDE_PHE_OEH_06.02_eng.pdf;jsessionid=202465FF8F388C645714FC1218341FF4?sequence=1.

14   Bandyopadhyay, A. Curr Pollution Rep (2016) 2: 203. ISSN 2198-6592, https://doi.org/10.1007/s40726-016-0039-z

15   "7 Million Premature Deaths Annually Linked to Air Pollution." *WHO*, World Health Organization, 25 Mar. 2014, www.who.int/mediacentre/news/releases/2014/air-pollution/en/.

16   Vohra, Karn, et al. "Global Mortality from Outdoor Fine Particle Pollution Generated by Fossil Fuel Combustion: Results from GEOS-Chem." *Environmental Research*, vol. 195, 2021, p. 110754., doi:10.1016/j.envres.2021.110754.

17   "Air Pollution in China Is Killing 4,000 People Every Day, a New Study Finds." *The Guardian*, Guardian News and Media, 14 Aug. 2015, www.theguardian.com/world/2015/aug/14/air-pollution-in-china-is-killing-4000-people-every-day-a-new-study-finds.

18   Ibid

19   "MSHA News Release: MSHA Issues Final Rule on Lowering Miners' Exposure to Respirable Coal Dust [04/23/2014]: U.S. Department of Labor." *U.S. Department of Labor Seal*, www.dol.gov/newsroom/releases/msha/msha20140669.

20   "Key Findings: State of the Air 2020." *American Lung Association*, www.stateoftheair.org/key-findings/#:~:text=The%20%22State%20of%20the%20Air%22%202020%20report%20shows,ozone,%20also%20known%20as%20%22smog,%22%20reached%20unhealthy%20levels.

21   Ibid

22   Ibid

23   "Does the Production of Natural Gas from Shales Cause Earthquakes? If so, How Are the Earthquakes Related to These Operations?" *USGS*, U.S. Department of the Interior, www.usgs.gov/faqs/does-production-natural-gas-shales-cause-earthquakes-if-so-how-are-earthquakes-related-these?qt-news_science_products=0#qt-news_science_products.

24   This figure is relative to a time period of a century. Since most methane released is "scrubbed" from the atmosphere within 10 or so years (while CO_2 may remain for a couple hundred years) its impact becomes less prevalent over longer time scales. Over a 20-year period, methane is 84-87 times more potent, but over the lifetime of CO_2 methane is "only" 21-25 times more potent.

25   Voiland, Adam. "Methane Matters." *NASA*, 8 Mar. 2016, earthobservatory.nasa.gov/features/MethaneMatters.

26   "Nuclear Power in France." *World Nuclear Association*, Oct. 2019, www.world-nuclear.org/information-library/country-profiles/countries-a-f/france.aspx.

27   Parry, Ian, et al. "Global Fossil Fuel Subsidies Remain Large: An Update Based on Country-Level Estimates." *International Monetary Fund (IMF)*, 2 May 2019, www.imf.org/en/Publications/WP/Issues/2019/05/02/Global-Fossil-Fuel-Subsidies-Remain-Large-An-Update-Based-on-Country-Level-Estimates-46509. ISBN/ISSN: 9781484393178/1018-5941

28   "Direct Federal Financial Interventions and Subsidies in Energy in Fiscal Year 2016." *U.S. Energy Information Administration (EIA)*, Apr. 2018, www.eia.gov/analysis/requests/subsidy/pdf/subsidy.pdf. (Page 9, pulled from table 3 "Quantified energy-specific subsidies and support by type, FY 2010, FY 2013, and FY 2016")

29   Ibid

30   IEA (2020), Global Energy Review 2020, IEA, Paris https://www.iea.org/reports/global-energy-review-2020

31   Ibid

32   Ibid

33   IEA (2020), Coal-Fired Power, IEA, Paris https://www.iea.org/reports/coal-fired-power

34   Ibid

35   Ibid (Pulled from graph "Coal capacity by type in the Sustainable Development Scenario, 2010-2030")

36   Ibid (Pulled from graph "Coal capacity by type in the Sustainable Development Scenario, 2010-2030")

37 IEA (2019), World Energy Outlook 2019, IEA, Paris https://www.iea.org/reports/world-energy-outlook-2019 (Pulled from graph "Coal demand by region and scenario, 2018-2040")

38 Ibid (Pulled from graph "Coal supply investment, 2010-2018")

39 "Environmental Impacts of Natural Gas." *Union of Concerned Scientists*, 19 June 2014, www.ucsusa.org/resources/environmental-impacts-natural-gas#:~:text=Natural%20gas%20emits%2050%20to%2060%20percent%20less,burned%20in%20today%E2%80%99s%20typical%20vehicle%20[%202%20].

40 "Staff Report to the Secretary on Electricity Markets and Reliability." *U.S. Department of Energy (DOE)*, Aug. 2017, www.energy.gov/sites/prod/files/2017/08/f36/Staff%20Report%20on%20Electricity%20Markets%20and%20Reliability_0.pdf. (Page 13)

41 IEA (2019), World Energy Outlook 2019, IEA, Paris https://www.iea.org/reports/world-energy-outlook-2019

42 IEA (2020), Natural Gas Information: Overview, IEA, Paris https://www.iea.org/reports/natural-gas-information-overview (Pulled from graph "World natural gas demand by region, 1973-2019")

43 IEA (2019), World Energy Outlook 2019, IEA, Paris https://www.iea.org/reports/world-energy-outlook-2019 (Pulled from graph "Gas production by region and scenario, 2018-2040")

44 IEA (2020), Oil Information: Overview, IEA, Paris https://www.iea.org/fuels-and-technologies/oil

45 IEA (2020), Global Energy Review 2020, IEA, Paris https://www.iea.org/reports/global-energy-review-2020 (Pulled from graph "Change in monthly oil demand in selected countries, 2020 relative to 2019")

46 Ibid

47 "Oil and Gas Industry Employment Growing Much Faster than Total Private Sector Employment." *U.S. Energy Information Administration (EIA)*, 8 Aug. 2013, www.eia.gov/todayinenergy/detail.php?id=12451.

48 "The 2019 U.S. Energy and Employment Report." *Energy Futures Initiative (EFI) and the National Association of State Energy Officials (NASEO)*, www.usenergyjobs.org/2019-report. (Page 51, figures 29 & 30)

49 Ibid (Page 51, figures 29 & 30)

50 Ibid (Page 101, table 34)

Chapter 19: The World We Pass On

1   "The Paris Agreement." UNFCCC, United Nations, 22 Apr. 2016, unfccc.int/process-and-meetings/the-paris-agreement/the-paris-agreement.

2   Buis, Alan. "A Degree of Concern: Why Global Temperatures Matter – Climate Change: Vital Signs of the Planet." *Global Climate Change*, NASA, 25 June 2019, climate.nasa.gov/news/2865/a-degree-of-concern-why-global-temperatures-matter/.

3   FAQ 3.1, Figure 1 (Chapter 3) from Hoegh-Guldberg, O., D. Jacob, M. Taylor, M. Bindi, S. Brown, I. Camilloni, A. Diedhiou, R. Djalante, K.L. Ebi, F. Engelbrecht, J. Guiot, Y. Hijioka, S. Mehrotra, A. Payne, S.I. Seneviratne, A. Thomas, R. Warren, and G. Zhou, 2018: Impacts of 1.5°C Global Warming on Natural and Human Systems. In: *Global Warming of 1.5°C. An IPCC Special Report on the impacts of global warming of 1.5°C above pre-industrial levels and related global greenhouse gas emission pathways, in the context of strengthening the global response to the threat of climate change, sustainable development, and efforts to eradicate poverty* [Masson-Delmotte, V., P. Zhai, H.-O. Pörtner, D. Roberts, J. Skea, P.R. Shukla, A. Pirani, W. Moufouma-Okia, C. Péan, R. Pidcock, S. Connors, J.B.R. Matthews, Y. Chen, X. Zhou, M.I. Gomis, E. Lonnoy, T. Maycock, M. Tignor, and T. Waterfield (eds.)]. In Press.

4   "Impact of 1 °C and 2 °C Global Warming." Chapter 2- "Mitigation pathways compatible with 1.5°C in the context of sustainable development", *IPCC*, United Nations, 8 Oct. 2018, www.ipcc.ch/sr15/chapter/chapter-2/.

5   "Fluorinated Greenhouse Gases." *European Commission*, ec.europa.eu/clima/policies/f-gas_en#:~:text=Hydrofluorocarbons%20(HFCs)%20are%20by%20far%20the%20most%20relevant,cut%20by%20two-thirds%20by%202030%20in%20the%20EU.

6   Harvey, Chelsea, and Nathaniel Gronewald. "Greenhouse Gas Emissions to Set New Record This Year, but Rate of Growth Shrinks." *Science*, E&E News, 4 Dec. 2019, www.sciencemag.org/news/2019/12/greenhouse-gas-emissions-year-set-new-record-rate-growth-shrinks.

7   "Is It Too Late to Prevent Climate Change?" *NASA-Global Climate Change*, NASA, climate.nasa.gov/faq/16/is-it-too-late-to-prevent-climate-change/.

8   "How Cold Was the Ice Age? Researchers Now Know." *University of Arizona News*, 26 Aug. 2020, news.arizona.edu/story/how-cold-was-ice-age-researchers-now-know.

9   Tierney, Jessica E., *et al.* "Glacial Cooling and Climate Sensitivity Revisited." *Nature*, vol. 584, no. 7822, 2020, pp. 569–573., doi:10.1038/s41586-020-2617-x.

10   Ray, Louis L. "The Great Ice Age." ISBN 0-16-036025-0, Page 2, *U.S. Geological Survey*, U.S. Department of the Interior, 29 Aug. 2012, pubs.usgs.gov/gip/ice_age/ice_age.pdf.

11   Zuckerman, Wendy. "Climate Change... the Apocalypse?" Science VS., 5:30 gimletmedia.com/shows/science-vs/brhoa4/climate-change-the-apocalypse.

12   "Carbon Dioxide Levels Hit Record Peak in May." *NOAA Research*, 4 June 2019, research.noaa.gov/article/ArtMID/587/ArticleID/2461/Carbon-dioxide-levels-hit-record-peak-in-May.

13   Hausfather, Zeke. "The Water Vapor Feedback." *Yale Climate Connections*, 4 Feb. 2008, yaleclimateconnections.org/2008/02/common-climate-misconceptions-the-water-vapor-feedback-2/.

14   "A Daily Record of Atmospheric Carbon Dioxide from Scripps Institution of Oceanography at UC San Diego." *The Keeling Curve*, U.S. Department of Energy, Jan. 2020, scripps.ucsd.edu/programs/keelingcurve/.

15   "Climate Action Summit 2019 Closing Press Release." *UN Climate Action Report*, United Nations, 23 Sept. 2019, www.un.org/en/climatechange/assets/pdf/CAS_closing_release.pdf.

16   "Global Coal Plant Tracker." *End Coal*, endcoal.org/global-coal-plant-tracker/.

17   Ibid

18   Lindsey, Rebecca, and LuAnn Dahlman. "Climate Change: Global Temperature." *National Oceanic and Atmospheric Administration (NOAA)*, 15 Mar. 2021, www.climate.gov/news-features/understanding-climate/climate-change-global-temperature. (See graph "History of global surface temperature since 1880")

19   "Thawing Permafrost Produces More Methane than Expected." *Phys.org*, Helmholtz Association of German Research Centres, 20 Mar. 2018, phys.org/news/2018-03-permafrost-methane.html.

20   Study Citation: Knoblauch, Christian, et al. "Methane Production as Key to the Greenhouse Gas Budget of Thawing Permafrost." *Nature Climate Change*, vol. 8, no. 4, 2018, pp. 309–312., doi:10.1038/s41558-018-0095-z.

21   Gray, Ellen. "Unexpected Future Boost of Methane Possible from Arctic Permafrost." *NASA*, 20 Aug. 2018, climate.nasa.gov/news/2785/unexpected-future-boost-of-methane-possible-from-arctic-permafrost/.

22   "Thermodynamics: Albedo." *National Snow and Ice Data Center*, 3 Apr. 2020, nsidc.org/cryosphere/seaice/processes/albedo.html.

23   "Arctic Sea Ice Minimum." *NASA*, climate.nasa.gov/vital-signs/arctic-sea-ice/.

24   "Ice Sheets." *NASA*, climate.nasa.gov/vital-signs/ice-sheets/.

25   Notz, Dirk, and SIMIP Community. "Arctic Sea Ice in CMIP6." *Geophysical Research Letters*, vol. 47, no. 10, 2020, doi:10.1029/2019gl086749.

26   Ogburn, Stephanie Paige. "More Arctic Methane Bubbles into Atmosphere." *Scientific American*, 25 Nov. 2013, www.scientificamerican.com/article/more-arctic-methane-bubbles-into-atmosphere/. This source has the figure listed as "17 tera-grams" in the text, which is equal to 17 million tons. Changed in the writing of this book for ease of reading.

27   NOAA National Centers for Environmental Information (NCEI) U.S. Billion-Dollar Weather and Climate Disasters (2021). https://www.ncdc.noaa.gov/billions/, DOI: 10.25921/stkw-7w73

28   Ibid

29   Ibid

30   Ibid

31   Kahn, Matthew, et al. "Long-Term Macroeconomic Effects of Climate Change: A Cross-Country Analysis." *National Beureu of Economic Research*, Aug. 2019, doi:10.3386/w26167.

32   Multi-Model Framework for Quantitative Sectoral Impacts Analysis: A Technical Report for the Fourth National Climate Assessment. U.S. Environmental Protection Agency, Washington, D.C. www.epa.gov/cira

33   (Also see) "Hearing: The Costs of Climate Change – Risks to the U.S. Economy and the Federal Budget." *House Budget Committee Democrats*, 6 July 2019, budget.house.gov/publications/report/hearing-costs-climate-change-risks-us-economy-and-federal-budget.

34   Muggah, Robert. "The World's Coastal Cities Are Going under. Here's How Some Are Fighting Back." *World Economic Forum*, 16 Jan. 2019, www.weforum.org/agenda/2019/01/the-world-s-coastal-cities-are-going-under-here-is-how-some-are-fighting-back/.

35   "EPA FACT SHEET: SOCIAL COST OF CARBON ." *U.S. Environmental Protection Agency (EPA)*, Dec. 2016, www.epa.gov/sites/production/files/2016-12/documents/social_cost_of_carbon_fact_sheet.pdf. (Page 4)

36   "CO2 Emissions by Year." *Worldometer*, www.worldometers.info/co2-emissions/co2-emissions-by-year/. (Pulled from table "Global Fossil Carbon Dioxide emissions by Year")

37   "Groundswell: Preparing for Internal Climate Migration." *World Bank*, 19 Mar. 2018, www.worldbank.org/en/news/infographic/2018/03/19/groundswell—-preparing-for-internal-climate-migration.

38   Evans, Gareth. "Europe's Migrant Crisis: The Year That Changed a Continent." *BBC News*, 31 Aug. 2020, www.bbc.com/news/world-europe-53925209.

39   Bassetti, Francesco. "Environmental Migrants: Up to 1 Billion by 2050." *Foresight*, CMCC, 22 May 2019, www.climateforesight.eu/migrations-inequalities/environmental-migrants-up-to-1-billion-by-2050/.

40   "Global Report on Internal Displacement 2019." *Internal Displacement Monitoring Centre (IDMC)*, www.internal-displacement.org/global-report/grid2019/.

41   Ibid

42   Ibid

43   Ibid

44   "Climate Change Raises Conflict Concerns." *United Nations Educational, Scientific and Cultural Organization (UNESCO)*, en.unesco.org/courier/2018-2/climate-change-raises-conflict-concerns.

45   Ibid

46   "Climate Change Recognized as 'Threat Multiplier', UN Security Council Debates Its Impact on Peace." *United Nations*, www.un.org/peacebuilding/fr/news/climate-change-recognized-%E2%80%98threat-multiplier%E2%80%99-un-security-council-debates-its-impact-peace.

47   Mehta, Aaron. "Climate Change Is Now a National Security Priority for the Pentagon." *Defense News*, 27 Jan. 2021, www.defensenews.com/pentagon/2021/01/27/climate-change-is-now-a-national-security-priority-for-the-pentagon/.

48   Ibid

49   Litchwagger, David. "Ocean Acidification." Edited by Jennifer Bennett, *Smithsonian Ocean*, National Oceanic and Atmospheric Administration, 20 June 2019, ocean.si.edu/ocean-life/invertebrates/ocean-acidification.

50   "Ocean Acidification." *National Oceanic and Atmospheric Administration (NOAA)*, Apr. 2020, www.noaa.gov/education/resource-collections/ocean-coasts/ocean-acidification.

51   "A Primer on PH." *National Oceanic and Atmospheric Administration (NOAA)*, pmel.noaa.gov/co2/story/A+primer+on+pH.

52   Ibid

53   Ibid

54   "Ocean Acidification." *Woods Hole Oceanographic Institution*, 6 Feb. 2019, www.whoi.edu/know-your-ocean/ocean-topics/ocean-chemistry/ocean-acidification/.

55   "How Much Fish Do We Consume? First Global Seafood Consumption Footprint Published." *EU Science Hub*, European Commission, 27 Sept. 2018, ec.europa.eu/jrc/en/news/how-much-fish-do-we-consume-first-global-seafood-consumption-footprint-published.

56   Ibid

57   Ibid

58   Ibid

59   Schmidtko, S., Stramma, L. & Visbeck, M. Decline in global oceanic oxygen content during the past five decades. *Nature* 542, 335–339 (2017). https://doi.org/10.1038/nature21399

60   Cart, Julie, and Judy Lin. "California Fires Are Getting Worse. What's Going On?" *LAist*, CalMatters, 28 Oct. 2019, laist.com/2019/10/28/california-fires-explained-why-they-are-getting-worse.php.

61   Dahl, Kristy. "5 Of California's 6 Largest Fires on Record Are Burning Now: The Astonishing 2020 Wildfire Season in Context." *Union of Concerned Scientists*, 6 Oct. 2020, blog.ucsusa.org/kristy-dahl/5-of-californias-6-largest-fires-on-record-are-burning-now-the-astonishing-2020-wildfire-season-in-context.

62   Including where I live, in south-central Pennsylvania. There was a noticeable haze for about 4 days from my home. Below is a *Los Angeles Times* article that does a good job at describing the extent of that event.

63   Money, Luke, and Richard Read. "Smoke from California Wildfires Reaches the East Coast and Europe." *Los Angeles Times*, 15 Sept. 2020, www.latimes.com/california/story/2020-09-15/smoke-california-wildfires-reaches-east-coast-europe.

64   "Australia Fires: A Visual Guide to the Bushfire Crisis." *BBC News*, BBC, 21 Jan. 2020, www.bbc.com/news/world-australia-50951043.

65   McSweeney, Robert. "Deforestation in the Tropics Affects Climate around the World, Study Finds." *Carbon Brief*, 18 Dec. 2014, www.carbonbrief.org/deforestation-in-the-tropics-affects-climate-around-the-world-study-finds.

66   "Wildland Fire Research: Health Effects Research." *EPA*, United States Environmental Protection Agency, 4 June 2019, www.epa.gov/air-research/wildland-fire-research-health-effects-research.

67   "Global Warming and Hurricanes." *Geophysical Fluid Dynamics Laboratory*, National Oceanic and Atmospheric Administration (NOAA), 23 Sept. 2020, www.gfdl.noaa.gov/global-warming-and-hurricanes/.

68   "What Percentage of the American Population Lives near the Coast?" *National Ocean Service*, National Oceanic and Atmospheric Administration (NOAA), oceanservice.noaa.gov/facts/population.html.

69   Thornell, Christina. "Why Jakarta Is Sinking." *Vox*, 22 Feb. 2021, www.vox.com/22295302/why-jakarta-sinking-flooding-colonialism.

70   Lindsey, Rebecca. "Climate Change: Global Sea Level." *National Oceanic and Atmospheric Administration (NOAA)*, 25 Jan. 2021, www.climate.gov/news-features/understanding-climate/climate-change-global-sea-level. NOAA Climate.gov graph, adapted from figure 8 of Sweet, W.V., Kopp, R.E., Weaver, C.P., Obeysekera, T., Horton, R.M., Thieler, E.R., and Zervas, C. (2017). Global and Regional Sea Level Rise Scenarios for the United States. NOAA Tech. Rep. NOS CO-OPS 083. National Oceanic and Atmospheric Administration, National Ocean

Service, Silver Spring, MD.75pp. [Online: https://tidesandcurrents.noaa.gov/publications/
techrpt83_Global_and_Regional_SLR_Scenarios_for_the_US_final.pdf]

71   Patel, Kasha. "Drought Persists in the U.S. Southwest." *NASA*,
earthobservatory.nasa.gov/images/144216/drought-persists-in-the-us-southwest#:~:text=
Persistent%20drought%20conditions%20have%20spread%20across%20the%20U.S.,to%20the
%20National%20Integrated%20Drought%20Information%20System%20.

72   "A Closer Look: Temperature and Drought in the Southwest." *U.S. Environmental Protection
Agency (EPA)*, 2 Nov. 2020, www.epa.gov/climate-indicators/southwest.

73   Welch, Craig. "Climate Change Helped Spark Syrian War, Study Says." *National Geographic*,
2 Mar. 2015, www.nationalgeographic.com/science/article/150302-syria-war-climate-change-
drought.

74   Study Citation: Kelley, Colin P., et al. "Climate Change in the Fertile Crescent and
Implications of the Recent Syrian Drought." *Proceedings of the National Academy of Sciences*, vol.
112, no. 11, 2015, pp. 3241–3246., doi:10.1073/pnas.1421533112.

75   "Afghan Drought 'Displacing More People than Taliban Conflict'." *BBC News*, 17 Oct.
2018, www.bbc.com/news/world-asia-45872897.

76   Medetsky, Anatoly, and Megan Durisin. "Russia Halts Wheat Exports, Deepening Fears of
Food Shortage." *Time*, 27 Apr. 2020, time.com/5827804/russia-wheat-food-shortage/.

77   "Is It Too Late to Prevent Climate Change?" *NASA-Global Climate Change*, NASA,
climate.nasa.gov/faq/16/is-it-too-late-to-prevent-climate-change/.

78   "World Simply 'Not on Track' to Slow Climate Change This Year: UN Weather Agency |
UN News." *United Nations*, United Nations, 29 Nov. 2018,
news.un.org/en/story/2018/11/1026981.

79   "World of Change: Global Temperatures." *NASA*, earthobservatory.nasa.gov/world-of-
change/global-temperatures.

80   Patz, J. A. *et al.* "Climate Change and Infectious Diseases." *World Health Organization
(WHO)*, www.who.int/globalchange/publications/climatechangechap6.pdf.

81   Ibid (Page 114)

82   Jordan, Rob. "How Does Climate Change Affect Disease?" *Stanford Earth*, 15 Mar. 2019,
earth.stanford.edu/news/how-does-climate-change-affect-disease#gs.x49pvd.

83   Goudarzi, Sara. "How a Warming Climate Could Affect the Spread of Diseases Similar to
COVID-19." *Scientific American*, 29 Apr. 2020, www.scientificamerican.com/article/how-a-
warming-climate-could-affect-the-spread-of-diseases-similar-to-covid-19/.

84   "Special Report on Climate Change and Land." *Intergovernmental Panel on Climate Change
(IPCC)*, www.ipcc.ch/srccl/. (See chapter 3 executive summary)

85   Ibid (See chapter 3 executive summary)

86   "Climate Change: Land Degradation and Desertification." *World Health Organization
(WHO)*, 26 Oct. 2020, www.who.int/news-room/q-a-detail/climate-change-land-degradation-
and-desertification.

87   Haner, Josh. *et al.* "Living in China's Expanding Deserts." *The New York Times*, www.nytimes.
com/interactive/2016/10/24/world/asia/living-in-chinas-expanding-deserts.html.

88   "Report on Effects of a Changing Climate to the Department of Defense." *Office of the Under
Secretary of Defense for Acquisition and Sustainment*, U.S. Department of Defense (DOD), Jan. 2019,
climateandsecurity.files.wordpress.com/2019/01/sec_335_ndaa-report_effects_of_a_changing_
climate_to_dod.pdf.

Conclusions: Overcoming The Great Filter

1   Dodd, Matthew S., et al. "Evidence for Early Life in Earth's Oldest Hydrothermal Vent Precipitates." *Nature*, vol. 543, no. 7643, 2017, pp. 60–64., doi:10.1038/nature21377. (Alternatively see) Ghosh, Pallab. "Earliest Evidence of Life on Earth 'Found'." *BBC News*, 1 Mar. 2017, www.bbc.com/news/science-environment-39117523.

2   Fleming, Sean. "Climate Change Helped Destroy These Four Ancient Civilisations." *World Economic Forum*, 29 Mar. 2019, www.weforum.org/agenda/2019/03/our-turn-next-a-brief-history-of-civilizations-that-fell-because-of-climate-change/.

3   "Venus Overview." *Solar System- NASA*, NASA, 2019, solarsystem.nasa.gov/planets/venus/overview/.

4   "Mercury Overview." *NASA*, 10 Apr. 2019, solarsystem.nasa.gov/planets/mercury/overview/.

5   "The Atmosphere: Getting a Handle on Carbon Dioxide – Climate Change: Vital Signs of the Planet." *NASA*, 16 Oct. 2019, climate.nasa.gov/news/2915/the-atmosphere-getting-a-handle-on-carbon-dioxide/.

Index

www.ingramcontent.com/pod-product-compliance
Lightning Source LLC
Chambersburg PA
CBHW070108260726
48658CB00001B/35